# The Fabric of Interface

# The Fabric of Interface

## Mobile Media, Design, and Gender

Stephen Monteiro

The MIT Press
Cambridge, Massachusetts
London, England

This book was set in Stone Sans and Stone Serif by Toppan Best-set Premedia Limited. Printed and bound in the United States of America.

Library of Congress Cataloging-in-Publication Data

Names: Monteiro, Stephen, author.
Title: The fabric of interface : mobile media, design, and gender / Stephen Monteiro.
Description: Cambridge, MA : The MIT Press, [2017] | Includes bibliographical references and index.
Identifiers: LCCN 2017009921 | ISBN 9780262037006 (hardcover : alk. paper)
Subjects: LCSH: User interfaces (Computer systems) | Mobile computing. | Technology--Sex differences. | Gender identity.
Classification: LCC QA76.9.U83 M656 2017 | DDC 004.167--dc23 LC record available at https://lccn.loc.gov/2017009921

10   9   8   7   6   5   4   3   2   1

For Manisha and Dhruv

# Contents

Acknowledgments    ix

**Introduction    1**
1  Woven Memory    23
2  Image Fabric    59
3  Piecework    85
4  Domestic Disturbances    115

Notes    145
Index    181

# Acknowledgments

This undertaking has benefited from the assistance of a number of individuals and institutions. It started while I was a visiting professor at McGill University, but most of the work was done as a faculty member at Concordia University in Montreal. I thank my colleagues in Art History and Communication Studies at McGill, and in Sociology and Anthropology and in Communication Studies at Concordia, for their encouragement, especially Charles Acland, Greg Nielsen, Ahmed Al-Rawi, Matthew Unger, Jonathan Sterne, and Kyle Stine. Colleagues at other institutions have also provided help along the way. They include Kathleen Chevalier, Jayson Harsin, Mark Hayward, Nicholas Mirzoeff, Thomas Ort, Nanna Verhoeff, and Gregory Zinman.

Portions of my research were presented at the Society of Cinema and Media Studies annual conference in Seattle in 2014, the McGill AHCS Research Forum and Media@McGill in Montreal in 2015, and the International Association for Visual Culture biennial conference in Boston in 2016. I thank the organizers and audiences of these events for the discussions and debates they generated around my work. Segments of chapter 2 contain research that first appeared in my article "Fit to Frame: Image and Edge in Contemporary Interfaces," in *Screen* in fall 2014. I am grateful to that journal's editors for permission to include the material here. Similarly, I would thank those who have allowed use of the photographs and drawings that appear as figures in the book.

I have benefited from the expert help of librarians and archivists at the Massachusetts Institute of Technology, the MITRE Corporation, the Computer History Museum, McGill University, and Concordia University. I appreciate their detailed and prompt responses to my various inquiries.

At the MIT Press, Doug Sery has been a great source of support throughout the process. His steadfast belief in the project helped keep it on track. Noah Springer has guided me through multiples steps of production, and Kathleen Caruso's keen copyediting has helped clarify many details and statements in the text. The Press's anonymous readers offered generous and useful suggestions for improving this work, and I have tried my best to heed them.

For much of their lives, my four grandparents—Martha Mae Leach Hodgins, William Franklin Hodgins, Maria Augusta Temudo Monteiro, and Francisco José Monteiro—worked in and around the textile mills in Valley Falls, Central Falls, Pawtucket, and Manville, Rhode Island. Though the weaving had stopped before I arrived, their experiences profoundly influenced my view of the world and their memory has accompanied me at every step in the process of writing this book. My parents, Lois Ann Hodgins Monteiro and George Monteiro, and my sisters, Katherine Monteiro and Emily Morelli, also have helped me understand the value of that legacy.

Manisha Iyer continues to be an endless source for my inspiration. Her research on Indian textiles and weaving has provided me with important perspective and insights that helped illuminate the specificities and limitations of my own research and claims. Dhruv Iyer Monteiro provides quite another perspective, one that often rewards new ways of thinking. It is to them—and all they have given me—that I dedicate this work.

# Introduction

Contemporary digital media appear to have little in common with those of only a generation ago. In addition to transformations in their content, infrastructure, and application, our interactions and physical engagement with the digital media object have changed fundamentally. Encounters with digital networks and media frequently occur through handheld, electronic devices that accompany us through the day, carried in our back pocket or handbag. We turn and tilt these small plastic or metal frames with our hands and arms. We stroke and tap their glass screens with our fingertips. Through these physical interactions with the object and its surface, we make things: images, links, sites, networks. Our ability to effectively and efficiently identify patterns and build connections in this bodily performance, to bind the material of networked digital culture in new ways—whether it be in a game or on a social media platform—may earn us money, points, credit, followers, or some other desired quantitative reward.

Digital media's most unassuming components and operations are not self-evident or neutral entities, but cultural artifacts forged from long-standing social and ideological forces. As manual dexterity, patternmaking, and linking have risen to the forefront of everyday digital practice, our media interactions have taken on traits common to textile and needlecraft culture. Our smartphones and tablets share much with the handloom, the needlepoint hoop, and the lap-sized quilting frame. Each of these represents a portable platform, upon which one can create patterns, images, and other potentially meaningful visual configurations. Historically, looms, hoops, and quilting frames have been tools of the home, but they also have served as a means for greater social interaction, as with the communal functions of pattern sharing, fabric exchange, and quilting bees. Likewise, with the advent of the networked platforms for personal data and information sharing known collectively as social media, our portable electronics have become tools for a variety of interactions with others through the digital material that we access and its relation to our everyday circulation through the social sphere.

How may a handheld screen function like a loom, visual data function like swatches of fabric, and tactile interfaces function like needlework? What can such affinities tell us about communicative technology's adaptation of popular cultural codes? How does this create new ways of thinking about digital media's relationship to labor, identity, space, and the senses? Such questions frame the perspective and scope of *The Fabric of Interface*. Through its sustained exploration of weaving, fabric manipulation, and needlecraft as fundamental to historical and contemporary digital frameworks and interfaces, this book identifies important connections between contemporary networked media and practices often construed as alien to media technologies. It contends that social distinctions and gender divisions are reflected not only in what is made and circulated on digital devices and networks—as has been argued elsewhere[1]—but also through the immaterial and material forms, structures, and requirements of these devices and networks as they play out in electronic and physical actions and exchanges.

In their study of digital interaction design, Jay Bolter and Diane Gromala assert: "If we only look *through* the interface, we cannot appreciate the ways it shapes our experience. ... If we cannot also step back and see the interface as a technical creation, then we are missing half of the experience that new digital media can offer."[2] Stepping back to examine the correlation of digital and textile performativity in haptic and visual interface is significant for two reasons, both of which have consequences far beyond digital interactivity. First, it brings to the surface elements of computing's historical dependency on textile design, its production methods, and its labor models. This story is buried in computing's material past and scattered across its global sites of hard- and software manufacture and assembly, where women regularly have been responsible for the manual labor of weaving memory, threading hardwired programs, and integrating circuits. Second, the reification of this relationship in contemporary interface design and user practices raises vital questions about the relationship between gender and bodily interface in mobile media at a moment when such technologies would seem to transcend the issue. When considering ways an iPhone might be gendered, for example, one may be prone to begin and end with obvious marketing maneuvers such as the introduction of a pink (or "rose gold," according to Apple) back cover. "Are you man enough for a pink iPhone 6s?" *Esquire* asked its readers when the cover was introduced, referring to the result as a "powder-puff smartphone."[3] Yet such overt gestures toward the most conventional methods of coding gender in the everyday imply that these devices and their functionality are otherwise gender-neutral platforms upon which such codes may be added. In fact, the availability of colored covers or the production and use of deliberately gender-specific apps or language represent diversions that effectively obscure far more pervasive, but less easily identifiable, gendered characteristics of mobile media.

The look, feel, and function of contemporary media devices and their supporting software derive from deep-seated patterns of cultural practice, social structuring, and technological hierarchizing. This reflects Lisa Gitelman's contention that media are "muddy" entities requiring consideration of how they are formed through social protocol as much as how they function technologically. "Media include a vast clutter of normative rules and default conditions, which gather and adhere like a nebulous array around a technological nucleus," she states.[4] The approach of this book differs from Gitelman's, however, in its emphasis of the technological nucleus itself as constructed from normative rules and default conditions. In other words, socially constructed rules and conditions not only form around a technology, but also contribute significantly to that technology's formation in the first place. Any medium, any technology, is already muddy when it comes out of the box.

This book explores the muddy roots of networked digital media's forms and practices in emphasizing their historical, cultural, and aesthetic dependency on gendered embodiment and labor forms. Making the link between sewing, weaving, and quilting and contemporary technologies gives access to new ways of conceptualizing hardware and software design, sensorial experience, and personal networked media practice. It contributes to an alternative historical narrative of digital interactivity—one centered on the relationship between gender and interface aesthetics. Recent changes in the screen as an interactive object and tool represent a critical turning point in this story, producing new physical and ideological relationships between user, device, and digital production. Any consideration of the material design and functioning of media—in this case, the hardware of portable touchscreens and wearables and the software that guides and brings meaning to our actions upon them—must be informed by these long-established gendered discourses of social differentiation and power imbalance that they reify and sustain.[5]

Producing such an alternative narrative sharpens our understanding of the ways contemporary digital media may represent new modes of social production and interaction, and in what ways they merely adapt and redeploy modes already embedded in the history of computing and digital communication. Specifically, this narrative challenges the perceived breaks between industrial (or commercial) computing and personal computing in the closing decades of the twentieth century, and personal computing and social media in the opening decades of this century. Bridging these shifts is digital culture's continued reliance on textile and needlecraft practices, techniques, and methods drawn from spheres alternately labeled as feminine, private, and domestic. In this way, qualities of intimacy and engagement seen as novel to today's touchscreen media devices are in fact attributes already present in earlier conditions of digital production, where manual gestures common to home handicrafts played a fundamental role in the manufacture of mainframes.

In attempting to uncover the little-explored material, ideological, and social links between networked, mobile media practices and textile and needlecraft culture, however, it is important to recognize clues that have long hovered near the surface. In particular, metaphors of textile and craft permeate the history of computing and communication networks. We do not have to reflect long before they spring to mind. Software developers and engineers "weave" code that includes "threads," such as bulk calls, to subroutines and threads of execution. Internet administrators and users "weave" the "web" with "threaded" discussions and by "linking" (a term for joining knitted fabrics). Data structures—from lists to trees—can be "zippered," and when files are compressed and uncompressed they are "zipped" and "unzipped." Problems in existing programs and their underlying software code are repaired with "patches" made from additional code. Digital images are "stitched" together or "quilted" by image-editing programs to produce larger images, such as landscape panoramas and game environments. All of this material is guided through the distributed network of the Internet via nodes of gridded circuits known as switch "fabrics."[6] The prevalence of textile metaphors conceptually marks digital practices in ways that distance them from other crafts. Although links might be drawn to scrapbooking, modeling, or other methods of object making, the imaginary of computing (from its earliest history, as we shall see) is fashioned from processes surrounding the making and assembling of cloth.[7] Despite the presence of so many examples of the language of textile crafts in discussions of computing and digital media, however, they rarely have been treated as emblematic of deeper connections between digital communications and what has been called "homecraft" or "women's work."

Apart from these linguistic clues, contemporary digital media's intersections with textiles and crafts may be most evident in the success of online crafting communities and marketplaces. Jack Bratich and Heidi Brush, two scholars who consider this trend, identify a convergence of craft and digital culture that they call "fabriculture." These observations are noteworthy and valuable, pointing to the way very old and very new media have harmonized and contributed to each other.[8] Another pair of scholars, Stella Minahan and Julie Wolfram Cox, adopt the term "Stitch'nBitch" to identify this trend, a name that emphasizes the relationship between collective needlecraft and interpersonal communication through and around digital networks. "*Stitch'nBitch* may be an example of a new way of connecting that is based on material production using traditional craft skills and yarns as well as the optical fibre and twisted pair cable used for telecommunications," they explain, representing "a local and global phenomenon in which production and consumption of gender, technology and society collide."[9] In addition to traditional forms of meeting to assemble textiles, virtual bees have sprung up, in which a quilter will send other members of the bee her fabric choices and suggestions for styling the block. Members will then sew the pieces into blocks and return them to her for final assembly.[10] Bratich mentions "peer-to-peer textiling" as a way to describe an evolving craft culture that combines online and off-line group communication, meetings, and exchanges of information.[11] While online marketplaces specializing in such handmade objects, including Etsy and Cargoh, also have thrived, digital networks have been the place to organize and promote craftivism, which seeks to expose and confront social injustice and inequality through craft.[12] Kirsty Robertson has argued, however, that craftivism's dependence on these global networks for organization and promotion may undermine the power of its anti-neoliberal message.[13]

Conversely, the design and operation of digital devices, interfaces, and networks have informed the aesthetics of crafts and hand production—from

**Figure 0.1**
Needlepoint grid produced by KnitPro 2.0 from a photo of a New England textile mill. Original photo by the author.

sculpture to rug hooking—in important ways. James Bridle has called this the "New Aesthetic," explaining that it "reproduces the structure and disposition of the network itself, as a form of critique."[14] A fitting instance of this intersection of network and craft aesthetics is KnitPro, a web application that allows users to convert GIF, JPEG, and PNG image files into gridded patterns for needlepoint and knitting.[15] Such apps convert the photographic image into a grid of tiles that already suggests a woven or embroidered surface (figure 0.1). Sites such as Sprite Stitch, a blog for "video game crafts," celebrate the visual correspondences between computer-generated imagery and textiles, with everything from cross-stitched Pokémon Kanto maps to crocheted Mario dolls.[16] These examples amount to an open dialogue between craft culture and digital media. Moving away from an explicit emphasis on textiles, Minecraft nevertheless would be an example of a popular game that joins the block construction of craft culture with the raster-and-pixel aesthetic of early screen-based computing.

Another area of overlap between textile culture and computing, one that departs from craft as described in the foregoing examples, is the development of wearable media. This includes eyewear and wristwear, although objects such as eyeglasses and wristwatches have, in fact, always been wearable media. Smart fabrics and e-textiles created by companies such as Intelligent Textiles are woven of a conductive warp and weft to produce clothing capable of cybernetic circuits.[17] Other textiles incorporate optical fibers. These have been used in warfare and medicine, but also sports and fashion.[18]

While some of the preceding examples enter into the present narrative at significant moments—particularly in discussions of gendered labor and the senses—for the most part they remain peripheral to its objective and argument. These initial layers of explicit intersection between digital media and textile culture—whether tied to technological innovation, the creation of motifs, or the migration of craft communities to electronic forums—can overshadow the more profound ways the culture of textiles has shaped digital media structures and practices. Beneath the surface is a deeper material, performative, and ideological intertwining of textile-based crafts and digital technologies, processes, and habits. This hidden layer of relations and correspondences exposes the larger social forces at work in the most basic elements and actions of everyday media use and their dependency on pre-digital notions of gendered labor and production. It is this layer that may help explain, for example, why mobile media users willingly contribute their labor and creative production to social media platforms at no cost, while the companies behind these platforms gain billions of dollars of profits from this situation.

## Objects and Processes

Writing about interfaces, Johanna Drucker asserts, "In a very real, practical sense we carry on most of our personal and professional business through interfaces. Knowing how interface structures our relation to knowledge and behavior is essential."[19] Invoked in diverse discussions and contexts, *interface* has become a twenty-first-century trope. According to Branden Hookway, it "describes a cultural moment as much as it does a specific relationship between human user and technological artifact."[20] The word gains traction through its connotations of advanced technology and contemporary communication forms, suggesting both immediacy and engagement. As a trope it represents and conveys important ideological meanings. Wendy Chun has stated that the digital object's interface is "a functional analog to ideology" in that such devices can formally produce false consciousness and represent an individual's imaginary relationship to the real. "The 'choices' operating systems offer limit the visible and the invisible, the imaginable and unimaginable," Chun claims, adding that "interfaces also produce users through benign interactions."[21] Similarly, Drucker points out that an interface "combines two ideological illusions in a single paradoxical identity: the predictability of a mechanized automaton and the myth of autonomous agency."[22] As such, the interface as an idea as well as a series of actions between human and machine in time and space remains an important site in contemporary culture—perhaps the most important site—for the function of power on, and through, the individual. In their collection, *Interface Criticism: Aesthetics Beyond Buttons*, Christian Ulrik Andersen and Søren Bro Pold explain that "interfaces can embed choices, conduct, languages, and ultimately values, worldviews and aesthetics into technical infrastructures. … Today we perceive our environment through interfaces."[23]

Interface as it is explored in this study refers first and foremost to its most common meaning for media today—the site of contact between people and portable networked devices, and the events that take place there. This is most frequently experienced through the bodily manipulation of a screen-object and consequent changes in the digitally produced effects manifested by the object at or around the point of contact. Of course, software operations and network flows that underlie and facilitate this superficial, material activity are integral to the process. The understanding of interface deployed here draws on theories articulated by Hookway and Alexander Galloway, both of whom study interface in ways that reach far beyond popular, contemporary understandings of the term, to consider how an interface may surpass these parameters while nevertheless containing them at its center.[24] Galloway's idea of the interface as a "threshold" (Hookway uses the term as well) related to the exercise of neoliberal ideologies is particularly germane to the claims made here.[25] The interface allows, encourages, and guides certain patterns of activity and production practiced in multiple contexts that further the breakdown between distinctions of work and leisure, for example, while quantifying and monetizing our most basic and necessary social interactions. In his exploration of the effects of interface, Galloway explains: "Interfaces are not simply objects or boundary points. … Interfaces are themselves the effects of other things, and thus tell the story of the larger forces that engender them."[26] Studying the effect of interface—and interface as effect—in regard to contemporary mobile communication lies at the heart of this book as well.

Hookway similarly considers the interface as "a relation with technology rather than as a technology in itself." For him, "the interface describes a boundary condition that is at the same time encountered and worked through toward some specific end."[27] This meeting of two otherwise distinct entities produces what he calls "augmentation," which is aimed at control and power. Augmentation could be called the event of interface, and the chapters that follow argue that control and power through this encounter extend well beyond the interface because augmentation draws upon existing social conditions and actions.[28] This points to an additional aspect of interface considered here, especially in chapter 4: interface as physical space organized to regulate exchanges between people, space that increasingly accommodates interactions between the body and digital technologies. Such a space could be a living room, bedroom, or kitchen, but it is also a lobby, a park, a classroom, and even the sidewalk. As Hookway observes, the interface is a contested zone, where social and material meet, one that "governs the production of sites and events."[29]

In considering the relationship between portable media devices, social hierarchies, and communication, the mobility of devices (and our access to their networks) produces significant changes in our understanding of the meaning and function of our physical spaces and their ties to the ideology of a (problematic) public sphere of face-to-face interaction and expression. According to Andersen and Pold, "The interface is a dominant cultural form providing a way to mediate between humans and machines and between culture and data, affecting the way we perceive cultural activities and perform them in public and private."[30] From home to office to street, ideas of private and public space, of workspace and leisure space, are challenged by the paradoxically constant, but intermittent, use of networked portable devices throughout the day. This embodied performance of the digital interface across vastly different social spaces has the potential to challenge engrained assumptions concerning the role of those spaces and—refracted through gender, race, and other social lenses—what sorts of expressions and actions they will permit.[31] This is at the heart of what Malcolm McCullough has called the "ambient commons," an intermingling of spaces, bodies, and networks that not only changes the perceptual parameters of everyday living, but also requires new approaches to understanding and assessing attention, distraction, and participation.[32]

While interface is a thoroughly modern and relatively uncontested concept, one developed in nineteenth-century masculine spheres of fluid dynamics and mechanical engineering (according to Hookway), gender is a much older idea that has received considerable scrutiny and challenge in contemporary thought. Gender can be considered in at least two ways significant to the claims of this book. First, it exists as a qualitative distinction that can be applied to a greater or lesser extent to any form of human activity, communication, or invention. This is much the way grammatical gender functions in many languages, marking nouns as masculine or feminine. Second, it can be considered to be the process of making distinctions and the ensuing consequences of these distinctions as they impact and form identity, agency, and the place of the body within the social sphere. Gender, as it is considered in *The Fabric of Interface*, develops from both of these possibilities, not only signifying cultural distinctions of masculinity and femininity and their sociopolitical applications for regulating public and private spheres, but also serving as the linchpin in this book's exploration of the relationship between material production and social action. The body is at the center of this operation much as it is in critical explorations of gender and sexuality such as Judith Butler's *Bodies That Matter*, insofar as the material expression of sexual difference and the materiality of the body are considered here as they intersect with contemporary media forms and structures. Butler speaks of a "process of materialization" in regard to sex and gender as social entities, one that "stabilizes over time to produce the effect of boundary, fixity, and surface we call matter."[33] Elsewhere, she has argued that "if a true gender is a fantasy instituted and inscribed on the surface of bodies, then it seems that genders can be neither true nor false, but are produced as the truth effects of a discourse of primary and stable identity."[34] The relationships between bodies and these truth effects have been the object of study of Elizabeth Grosz, who has endeavored to tease out the connections across gender, body, and space. Building on feminist scholarship on the political role of the gendered body from Simone de Beauvoir to Iris Marion Young, Grosz asserts that "patriarchal oppression … justifies itself, at least in part, by connecting women much more closely than men to the body." In this way, "women's corporeal specificity is used to explain and justify the different (read: unequal) social positions and cognitive abilities of the two sexes."[35]

The argument elaborated here builds on existing theorizations of the relationship between the performativity of the body and the circulation of gender by describing the interface's relationship to gender at the levels of material design, physical interaction, and software coding. The way gender is regulated through these, how it is represented and performed, are among the most prominent and significant of the discursive effects raised by Butler and explored by Grosz. Indeed, contemporary hardware and software's frequent need for regular bodily intervention to function effectively would suggest that the human body itself may act as their interface within this context. This book therefore extends the important theoretical discussions of technology, gender, race, and identity already undertaken by scholars including Wendy Hui Kyong Chun, Lisa Nakamura, Donna Haraway, N. Katherine Hayles, and others, which—invoked in the pages that follow—have important implications for gender studies.

It is also imperative to point out that there is a significant racial component to the gendering of both craft culture and media's materiality and processes. For example, quilting took on particular importance among African American women in the nineteenth century as a means to make inexpensive blankets by sewing together scraps from worn garments as well as a means of building and preserving family and historical narratives in the face of politically enforced low literacy rates. Similarly, industrial assembly in the electronics industry has relied on the work of women of color, from the first digital computer with memory, which was hand-strung by an African-American woman technician in MIT's Lincoln Laboratory, to the latest smartphones put together by women in assembly houses scattered from China to Malaysia. While not the focus of this book, the role of race is considered at key points, building on the important work already done by Chun and Nakamura, among others.

### Existing Threads

Mobility and tactility are central components of the processes to be described and analyzed here. The proliferation of small, networked, touchscreen devices has produced dramatic changes not only in how we use devices and networks, but also where, when, and why. As recently as 1998, for example, computer scientist David Gelernter described a very different environment in musing on digital media design:

> Portable computers are easy to carry around, but it's likely to be quite a while before you can stuff one in your briefcase or coat pocket as nonchalantly as you can a sheaf of papers. ... You can work with papers when you are sprawled on a sofa, sitting on a beach, riding a subway, having coffee at a lunch counter, lying in bed unable to sleep at three in the morning, meeting with a dozen colleagues at a conference table and ostensibly examining the budget report. ... Yes, you *could* use a portable computer in all these circumstances, but it's a pain and few people do. Computerized files inevitably give something up to paper ones in portability, and that is likely to remain true for some time.[36]

Two decades into the twenty-first century, Gelernter's view appears comically myopic. Of course, this eventually becomes the case with most writing about technology, but it is worth noting that the Internet, cell phones, and handheld computing devices such as Apple's Newton and Palm's PalmPilot already were commercially available and in use when Gelernter made his remarks. His description demonstrates not only the difficulty of imagining large-scale portable networked computing in the late 1990s, but also the difficulty of conceiving of such computing except in terms of work and business (e.g., meetings and budget reports). By contrast, today's portable digital devices and their social media applications represent a brave new world.

While contemporary mobile media and their interfaces have become basic to everyday life in many parts of the world, critical inquiries into the social and historical origins of their formal logic and material attributes have lagged. Studies often privilege instead the transformative effects of their use. Though there have been some important exceptions—including work by Adrian Mackenzie, Tung-Hui Hu, Jonathan Sterne, and Matthew Fuller and Andrew Goffey, in addition to that of Chun, Galloway, Gitelman, and Hookway—many studies emphasize what digital objects and interfaces facilitate or produce (e.g., networks, data, social systems) rather than scrutinize their appearance, construction, and formatting.[37] Unlike the outlook of this book, existing research into the form and function of mobile media, software, and interface often also emphasizes movement and locativity—whether physical or imagined—as guiding properties tied to navigational, spatial, topological, and cartographic conceptualizations of interface and data exchange.[38] This bias toward structures of space and motion can reinforce the primacy of visual processes and effects traditionally coded as masculine—travel, exploration, route-finding, mapmaking, systems-building—in the development of these technologies and their use. Mobile media design researchers Frank Bentley and Edward Barrett, for example, claim that two of the three principal features of a mobile device are "always-available connection to others" and "sensing and capturing environments" within which the device circulates.[39] Despite the success of GPS and location-based social media such as Tinder, Grindr, and Foursquare, mobile media are never always mobile and much of what is done on them is not dependent on locativity. As portable media that often involve long periods of stationary user engagement, they are as much—or indeed more—about sedentary and intricate or repetitive tasks, as analogous to the handloom and quilting frame as they are to the compass, lens, or transmitter for visualizing, interacting with, or traversing space. For every minute a device is used to understand or interact with one's environment, there are likely to be many more where it is employed in watching videos, playing games, updating social media accounts, shopping, and reading the news.[40]

This book similarly challenges competing narratives of networked media use that situate the manipulation and recombination of digital data within a pedigree of modernist avant-garde remix practices such as collage, assemblage, and montage.[41] For example, in his contribution to *The New Media Reader* (an early and important collection on digital culture), Lev Manovich identifies in that book's contents a "notion of parallel developments in modern art and in computing."[42] He prescribes a list of propositions for new media that includes "new media as the encoding of modernist avant-garde" and "new media as parallel articulation of similar ideas in post-WWII art and modern computing." Accordingly, Manovich claims, "new media indeed represents the new avant-garde, and its innovations are at least as radical as the formal innovations of the 1920s," yet this new avant-garde is also post-media since it "is no longer concerned with seeing or representing the world in new ways but rather with accessing and using in new ways previously accumulated media."[43]

More recently, Eduardo Navas has taken a similar stance in his theory of digital culture as remix. For Navas, remixing describes common operations of creative expression emerging at the turn of this century, such as music sampling, mash-ups, and photo memes. He ties these to the avant-garde "as a cultural example in which remixing is at play explicitly on two layers that [he] define[s] as the framework of culture"—namely, the introduction of material and the process of producing value: "Social media relies on the framework of culture to develop a new type of economy. … Historically, creative production appears to resist established patterns of production, but eventually is sublated by cultural economies and becomes vital to capital as a whole."[44] While Navas's views on the absorption of culture resonate with the claims and examples in this book, his comparison of avant-garde tactics and everyday personal communication—as with Manovich's—raises more questions than it answers.

The problem with these formulations is twofold. First, they would suggest that the everyday user activity of putting together materials is guided by a similar, conscious effort to subvert existing power structures, even if this work is entirely inscribed within, and only possible through, the continued functioning of the technological manifestations of those structures. Although such subversion can be the object in specific cases, the comparison distorts the nature and aim of most networked assemblage activity on mobile devices (for example, using filters on social media photos), where assembly emerges as a conventional and conformist practice. Second, these formulations find novelty in processes of reuse, modification, and compositing that in many ways reflect and follow longstanding domestic practices of craft culture and handicraft's reliance on reusable and recyclable materials more closely than they do the countercultural aesthetic tactics of avant-garde movements. While sharing a belief in contemporary media's reliance on sampling and mixing, this book highlights this digital flexibility's compelling similarities to the "lower" (yet far more widely practiced, particularly among women) popular arts that constitute textile and needlecraft culture. Although they come closer to this book's interest in the digital's relationship to sedentary, absorptive material practices, references to the avant-garde—sometimes inflected with industrial overtones—perpetuate masculinizing narratives of technology and practice through their emphasis on a modernist history of art, photography, and film that was partially constructed on a distinction between elite culture and the history and aesthetics of craft.[45] While avant-garde movements may have contributed to some aspects of network culture, particularly among its earliest practitioners, far more familiar, popular cultural practices lie behind much of today's networked, digital interactivity.

In addition to the examples provided here of Bratich and Brush and Minahan and Cox, other researchers have pointed out links between digital culture and craft practices on occasion. This was particularly so in the 1990s, as the rise of the Internet produced any number of software practices diverging from previous norms. Asking "What kind of patroness would suit this virtual world of the net and the web?" Eric McLuhan nominated the weavers of Greek mythology, Arachne and Penelope, to represent the decentralized scope and endless production of digitally networked living.[46] Malcolm McCullough found a close relationship between interactive computing and craft in his *Abstracting Craft: The Practiced Digital Hand*. "In digital production, craft refers to the condition where people apply standard technological means to unanticipated or indescribable ends," he explains. "Works of computer animation, geometric modeling, and spatial databases get 'crafted' when experts use limited software capacities resourcefully, imaginatively, and in compensation for the inadequacies of prepackaged, hard-coded operations."[47] While McCullough makes important claims about the role of the hand—typically grasping a mouse, at the time he wrote—his idea of craft is based on the tradition of craftsmanship and the craftsman. It is of the workshop and the guild, not of the woman occupied with the domestic labor of the home. His contribution serves as a reminder that "craft" is a notoriously ambivalent word, applicable to diverse and sometimes opposite situations.[48] In this book, craft is not the historically superior cultural form of the craftsman, but emphatically the "inferior" form of production that has been called craft because it has been deemed unskilled, intellectually vacant, or of little economic value.

Working during the same period, Brenda Danet drew a correspondence between the techniques of making and sharing images in Internet Relay Chat (IRC) and crafts, explaining that the synchronous, collaborative aspect of chat rooms aligns with ideas of folk art. Danet likened IRC images built from typed characters to the stitch-based crafts of embroidery and needlepoint, not only because of their appearance, but also because of the hours it would take constructing an image from individual keystrokes. The collective sharing and display of these images in chat rooms borrows from techniques of quiltmaking and the social process of the quilting bee.[49] "The analogy with quilting is especially apt because quilting is more often a social activity, at least in part … and because [IRC] images often resembled those of quilts," Danet states.[50]

Sadie Plant's *Zeroes + Ones: Digital Women and the New Technoculture* and Jack Bratich's "The Digital Touch: Craft-Work as Immaterial Labour and Ontological Accumulation," in addition to Bratich and Brush's "Fabricating Activism: Craft-Work, Popular Culture, Gender" and Minahan and Cox's "Stitch'nBitch: Cyberfeminism, a Third Place and the New Materiality," are perhaps the most compelling texts in regard to textile crafts and digital culture, however. Each of these engages the relationship between digital culture and textiles and handicrafts as it pertains to matters of gender, power, and agency.[51] All four are treated in greater detail in the pages that follow, but it is worth pointing out now how each diverges from the present study in significant ways. Plant's work lays down a feminist interpretation of computing as a weaving process, emphasizing the role of Augusta Ada King, Countess of Lovelace, in the early development of digital and information theory. It is significant as a theoretical exploration that undoes much of the historical record and "common sense" that had surrounded digital culture and masculinity. Yet it also extends that exploration in ways that lead away from materialist readings of technology. Also, because it was published in 1997, many of its assertions are less applicable to the tactile, networked mobile media and digital image environment at the center of this book. Plant's claims therefore are more useful in consideration of the history of computing, but less vital when measured against the major cultural changes in digital media use in this century. The work of Bratich and Brush and Minahan and Cox, on the other hand, emerges from these changes. They are on the mark in their comparisons of textile and digital practices, particularly in linking immaterial digital labor to an underlying communicative, immaterial component of textile-based craftwork. However, they do not bring this to bear on the mechanics of media use in any sustained way, nor do they address the larger question of mobile media's relationship to gender as embodied practice, as is the case here. In their focus on the historical and theoretical components of the relationships among digital culture, gender, and textile crafts, other areas—such as material practices and design issues—remain underexplored.

## Pattern of the Book

Four points orient the structure and progression of *The Fabric of Interface*. They roughly correspond to its four chapters. The first point is that the history of computing has always relied on the sort of gendered labor that this book associates with, and identifies within, contemporary interfaces. The second is that understandings of the digital image as being of a different nature than images of the past have contributed to particularly gendered aspects of contemporary interfaces. This is significant in our increasing reliance on the image as a communicative and social tool. The third point is that the physical gestures and networked repercussions of contemporary, image-based interfaces mimic those of gendered textile work, suggesting deeper social and economic correspondences between often exploitative needle-based assembly and networked, immaterial labor. The last point is that ingrained beliefs in gender differences surrounding the senses produce value judgments about technology use and what constitutes a successful interface. In particular, these beliefs contribute to negative connotations of haptic interactivity. Let us consider how these points unfold in their respective chapters.

Chapter 1 establishes historical links across textiles, gender, and computing by tracing the role of needlecraft and textile techniques in the production of hardware and software from their conception in the nineteenth century through twentieth-first-century globalized electronics industries. It explains how Joseph Marie Jacquard's industrial loom apparatus of the early nineteenth century relied on complex weaving sequences—often to produce fabrics bearing elaborate images—that were stored on punched cards read by the loom's rods. Serving as a primitive form of computing memory, the stack of attached cards need only have been loaded into the mechanism to produce the textile. The introduction of such programs allowed quick, accurate, and repeated production of intricate weaves, obviating the need for highly paid, skilled weavers and opening the door to the increased presence of women as loom tenders, a trend noted by computing pioneer Charles Babbage. In 1843, Ada Lovelace—often considered the first computer programmer—wrote that Babbage's own Analytical Engine "weaves algebraic patterns just as the Jacquard-loom weaves flowers and leaves."[52] Babbage saw the punched-card programs of the loom as the key to programming computing machines, and articulated his invention through the language of textile manufacture.

While punched cards had facilitated automated weaving and the origins of computing, more sophisticated twentieth-century computers relied on hand weaving to create the hardware behind their versatility. At MIT's Lincoln Laboratory and later at IBM plants in North America and Asia, women produced computer memory by interlacing metal filaments through small magnetic rings called "cores." These fabrics of memory followed the principles of on-loom bead weaving by locking each core into the grid of filaments through multiple threading, in this case of driving, sensing, and inhibiting wires. A similar process was behind the production of the Apollo space program's memory, when NASA hired retired women textile workers to weave rope memory for space capsule navigation systems. Even in the shift to total plant production of semiconductor-based memory in the 1960s—first in the western United States, then Asia—women remained responsible for hardware assembly, soldering integrated circuit boards and microchips. Assembly houses in Southeast Asia have relied primarily on low-paid female hand labor often drawn from textile manufacturing or trained in domestic handicraft, raising important questions about gender and manual labor (as well as race) in "pre-tactile" digital culture.

Chapter 2 examines the influence of textile culture on the defining role that images play in contemporary digital media. It considers textual and visual discourses around the digital image that position it as a malleable fabric across its production, access, and use. Beginning with the construction of the image itself, this chapter draws analogies between the screen device and the handloom or quilting frame as the physical structure upon which images are manipulated individually and in patterned series. Each digital image's material instantiation and visual presence exists as a raster grid of picture elements (pixels) displayed on the screen's patterned weave of diodes. Stored in long strands of binary code, the image is only visible when these strands pass through a microprocessor. Accordingly, every digital image may be understood as a performance of weaving that ends only when the file is closed or the screen is darkened. In multi-image displays, software algorithms regularly arrange images in vertical and horizontal patterns according to their shape and content, much as a quilter arranges blocks on a frame.

These characteristics of digital imagery set it in opposition to understandings of analog photographic images developed in nineteenth-century positivist—and twentieth-century modernist—discourses. Yet the early history of photography closely links the medium's processes of image making to computational thinking through the material and metaphor of the textile. Exploring that history reveals that negative-positive photography inventor William Henry Fox Talbot's interest in fabrics as a primary photographic subject able to demonstrate his invention's properties brought photography into the gridlike structure and cultural logic of digital image production and display at the medium's earliest moments.

Consideration of the basics of digital image–processing systems reveals the continuing, explicit use of concepts of textile assemblage in the manipulation of files to create an integrated visual fabric. "Image stitching" and "image quilting" employ algorithms to produce "seamless" images, whether by joining fragments into a whole (as in a series of photos of partial views of a landscape) or by taking a single image pattern, multiplying it, then assembling these pieces of visual material into a larger image. Such processes—essential to building virtual visual environments—recognize the photograph as a patterned swatch that, when combined with other patterned swatches in particular ways, can produce further patterns prompting differing visual and perceptual experiences. These developments mark a gendered discursive shift in the conversion from film-based to digital imagery as the language of mechanical assembly common to film production—splicing and compositing—is replaced by the feminized language of sewing and quilting.

Recent interfaces for accessing stored digital images, such as infinite scroll, extend the logic of swatch integration into networked processes of mobile media by arranging multiple images into a patchwork quilt. Infinite scroll—commonly found in operations such as search engine results—recalls the associative logic of nineteenth-century "album quilts" as it visualizes results aggregated from distributed networks in an ever-lengthening whole as users scroll down a display. "Stitched" together as it is extended by the user, this length of interlocking images can be restitched in turn into new layouts simply by modifying search parameters.

Chapter 3 considers the relationship between the digital labor of networked, mobile media assemblage practices (such as liking, linking, and tagging) and earlier forms of collective, "mobile" production such as quilting bees and textile industry piecework. It explores similarities between the interactive, integrative practices of contemporary digital devices and their networked users, on the one hand, and the aims and mechanics of sewing and needlework, on the other. Correspondences between these are embodied in the multiple, basic operations of contemporary interfaces, social media platforms, and casual game apps, where matching and arranging material is a common activity. Such processes of contemporary digital culture parallel the quilting bee or other social groups created out of networked patterns of labor, and are marketed to users through a rhetoric of "groups" or "communities" rather than "organizations" or "networks." Immaterial, affective digital labor blurs the line between work and leisure in ways common to textile and handicraft culture, which often takes place amid other domestic responsibilities and tasks, such as cooking and childcare. In contrast to the algorithmically generated patterns studied in chapter 2, these activities bear a closer resemblance to the gendered labor structures presented in chapter 1 and represent a new phase of the "home economy" of gendered digital production. Their affinities to women's work permit intermittent and frequent interactions with the device and network that—despite their brevity—are monetized by network platforms with little or no remuneration for users. Yet they also lead to critiques of social media and casual game use as little more than frivolous, unproductive distractions.

Chapter 4 incorporates the findings of the preceding chapters into an assessment of the role of gender and sensory distinctions in contemporary mobile media. It considers the consequences of the parallels established across textile, needlecraft, and digital culture in relation to wider gender structures around work, embodiment, and visuality. Beginning with two historical binaries of mind and body, and sight and touch, as these have been deployed in the ideological exercise of gender as a sociopolitical tool, it demonstrates how both have influenced understandings of mobile touchscreen media and their use in diverse social contexts. Handheld devices with tactile interfaces connote intimacy, moving from the desktop to the lap or pocket, from the office to the living room or bedroom. Patchwork, haptic interface techniques of *bricolage* accompany these shifts. This convergence of intimacy and handiwork at the screen interface, coupled with the textile aesthetics mentioned earlier, genders networked culture and activity in unexpected ways. By obliging touchscreen users to direct their eyes downward to the screen, for example, the proliferation of these activities in public spaces and collective societal contexts—from the coffee shop to the street—can challenge the forward, upright, masculinized gaze that has historically constructed and dominated these environments. Construed as distraction and disruption, these activities seem to undermine the idealized "open," face-to-face contact of the public sphere. Amid this potential threat, virtual reality (VR) systems, and the totalizing view they promise, have reemerged as an area of consumer technology development after lying relatively dormant for nearly two decades. This chapter considers hands-free, vision-based augmented reality (AR) and VR systems such as Google Glass, Google Cardboard, and Oculus Rift—and the rhetoric surrounding them of a return of agency—as a presumed next step in networked media. While these technologies may estrange users from their immediate physical and social environments, they nevertheless offer the illusion of a retrieved, unimpeded gaze, where the interface seems to disappear. What such developments illustrate is the extent to which gender may enter—explicitly or not—into the design and use of devices and software, regardless of the content these technologies may contain or convey. This realization not only underlines the importance of investigating computing's extensive ties to activities of textile and needlecraft culture historically gendered as feminine, but also demonstrates the larger need to reconsider the social consequences of hardware and software design itself.

# 1  Woven Memory

The texture of mobile media appears to be smooth and unmodulated. Clean, firm glass screens facilitate the quick, even movements of fingertips. Yet this seamless surface hides textured patterns of lines, channels, and filaments, some of them crafted by the hands of unseen workers in different parts of the world. Indeed, a smartphone or tablet can be at once a woven object, a sewn object, and a handmade object, from the fine grid of sensor filaments placed just under the glass to the soldered circuit boards deep inside the casing. And this condition is nothing new, but only the most recent example in computing and electronics' long reliance on textile culture and needlecraft to make its machines and networks function. Weaving, textile assembly, and digital culture share a long history. Some of it is well known, some of it is scarcely documented, and some of it undoubtedly has been lost. While this history has been left in the shadows of contemporary media culture, it should not come as a surprise when we think about the nature of textile production or computing. As different as fabric patterns and software may seem, they both represent highly structured systems of calculation. Just as a laptop, tablet, or smartphone converts the binary code of ones and zeroes of software and data—conveyed by switching electrical currents on and off—into meaningful images for us, a loom transforms rows of perpendicularly stretched thread into meaningful patterns through similar up/down, on/off settings. When studied closely, the overlap in computing and textile cultures is substantial. Their convoluted relationship surfaces in the forms and practices of early computers as well as those of contemporary networked mobile media. Today's digital devices and software, particularly those involving haptic, image-based interfaces, draw heavily on the cultural characteristics of textiles and needlecraft. This propensity has consequences for our understanding of the societal role digital media play not only in the content they aggregate and circulate, but also in their appearance and performance in the spaces of the everyday. This relationship, moreover, has been critical for the theorization and conception of the logic and mechanics of modern computing from its beginnings two centuries ago.

### Algebraic Patterns

So many histories of modern computing begin in the nineteenth century with the Jacquard silk-weaving loom apparatus and its influence on Charles Babbage's Analytical Engine that the story stands as the originary mythology of digital culture.[1] However, this story is often conveyed simply as an inspirational tale of one technology impacting the design of another. The relationship between Babbage's ideas and textile culture are more complex, impacted not only by Joseph Marie Jacquard's device, but also Augusta Ada King, Countess of Lovelace's theorizations of programming and Babbage's views concerning production organization and labor in the textile industry. All of these are considered in this chapter as evidence of textile culture's fundamental role in the conceptualization of computing.

Although commonly called the Jacquard "loom," the apparatus invented by the French textile manufacturer and engineer in 1801 was more precisely a treadle-operated shedding mechanism that could be mounted over a drawloom (figure 1.1). The device automatically manipulated the vertical warp threads of the loom between each passage of the horizontal weft thread by means of a shuttle, the basic action of loom weaving (figure 1.2). The apparatus's automatic adjustment of threads was regulated by a system of pasteboard cards. Holes were punched on a card to correspond to a particular alignment of the warp rods—attached to the vertical threads of the loom—during a single horizontal passage of the shuttle. As each card passed under levers connected to the rods, those levers that lined up with the card's punched holes would fall, shifting the corresponding rods and their threads into the raised "on" position (figure 1.3). All other rods would remain in a default "off" position. The cards would be strung into chains—of hundreds or even thousands—to produce the "program" for the production of a specific cloth design.[2] In this configuration Babbage, an English mathematician and engineer, would later see the means of programmable, mechanized calculation.

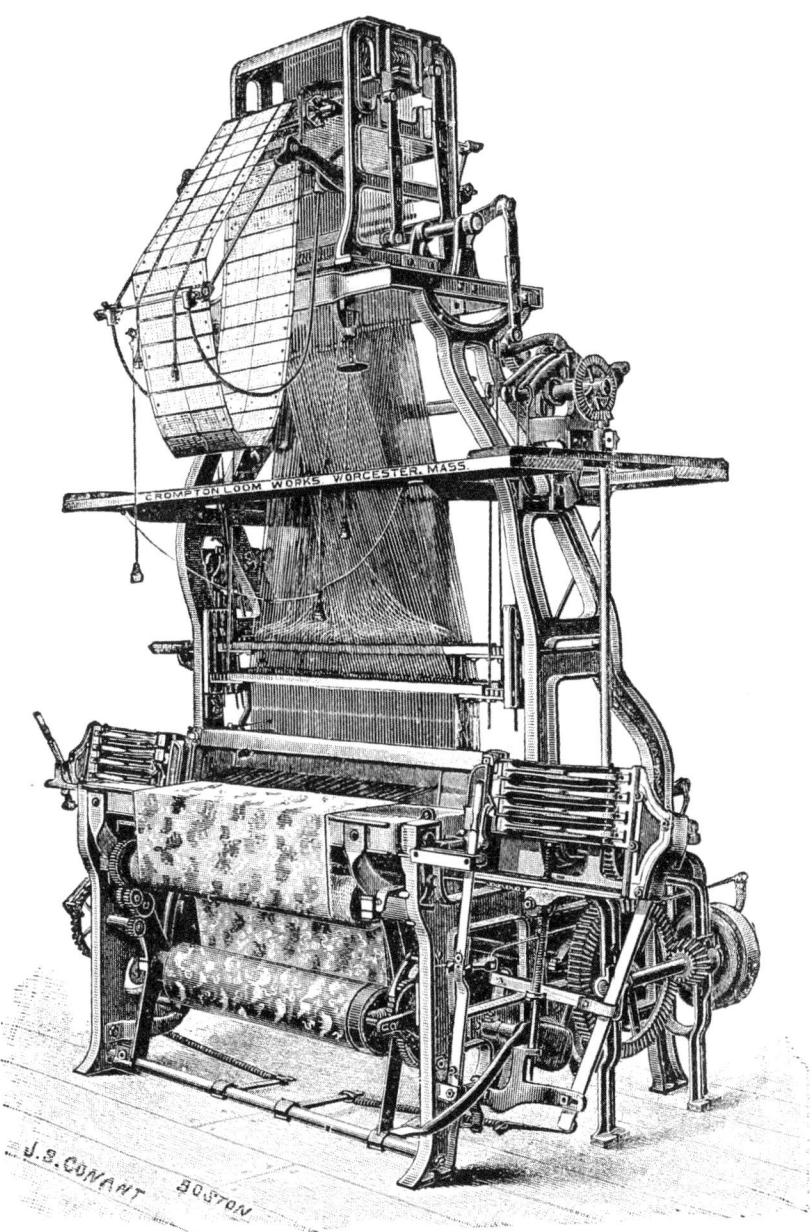

**Figure 1.1**
Loom equipped with a Jacquard mechanism, c. 1870. The punched cards used to adjust the warp threads can be seen extending from the upper left. Engraving by J. S. Conant Company, Boston.

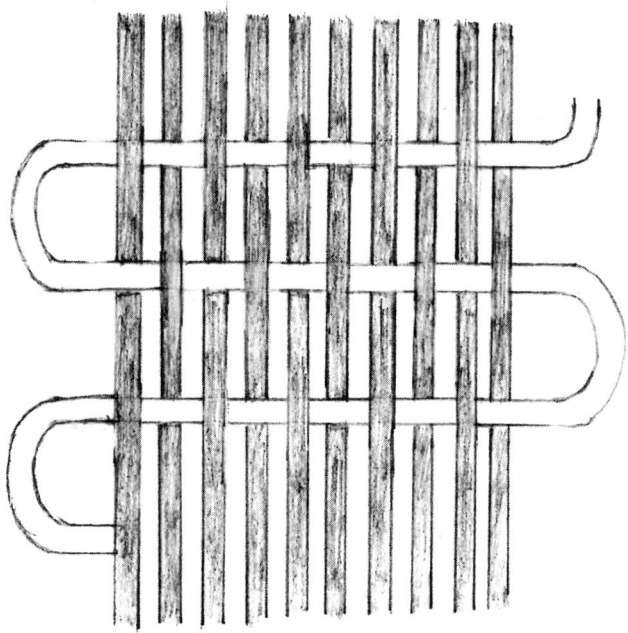

**Figure 1.2**
Diagram of a plain weave technique with a light weft thread passing horizontally through a line of darker warp threads that would be mounted on a loom frame. Drawing by Manisha Iyer.

Jacquard's apparatus permitted complex warp sequencing for elaborate patterns and images to be woven into silk textiles. It obviated the need for a human weaver to understand and execute these sequences. It replaced outright the drawboy (sometimes a drawgirl), who would sit above or beside the loom to manipulate weighted cords attached to warp threads according to the weaver's instructions.[3] Once a set of punched cards had been created to produce a specific image, motif, or pattern with this apparatus, the set could be stored, retrieved, and run at any time. Ostensibly, these programmed weaving combinations could be executed on any suitable drawloom, regardless of the talent or dedication of the operator. Once the first card was fed into the mechanism, the rest followed automatically. Like the more familiar example of the punched musical scroll in a player piano, Jacquard's card-based manufacturing system eliminated the need for real-time human calculations in the production of the piece. The passage of

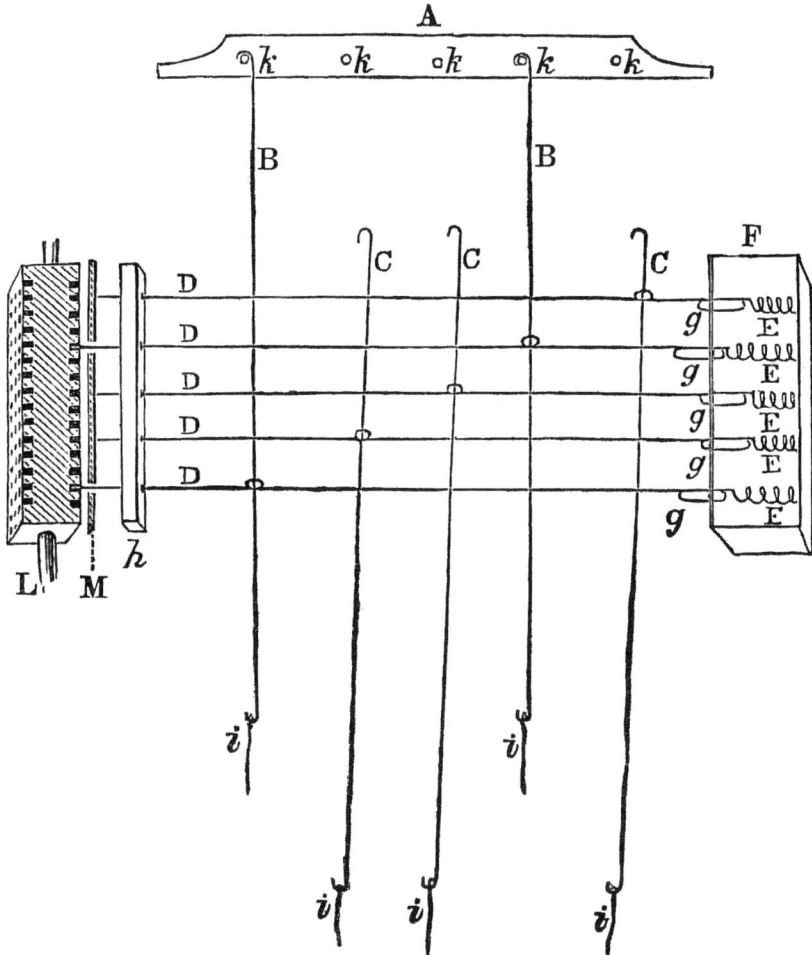

**Figure 1.3**

Schema of a Jacquard mechanism. The punched cards pass over the rotating block (L), causing rods (D) to either remain in place or drop, thereby allowing the warp thread hooks to either catch (B) or miss (C) the lifting bar (A) during a passage of the weft. From *Chambers's Encyclopædia: A Dictionary of Universal Knowledge* (London: Chambers, 1908).

hundreds of cards could render fabrics bearing intricate patterns and even entire images of a resolution comparable to that found in engraving or lithography (figure 1.4). Jacquard mechanisms are still fixed atop industrial looms today for the production of figured textiles. Contemporary examples rely on electronics rather than cards, however, and allow designers to weave directly from whatever pattern they create on a monitor through the use of "weave software," much like what-you-see-is-what-you-get desktop printing.[4]

Just as Jacquard's apparatus would have fundamental consequences for the architecture and functioning of digital computers, it profoundly influenced industrial production and labor systems, making it a key instrument in the Industrial Revolution as well as in the relationship between computing and social structures. While these aspects of Jacquard's invention and its history have received attention individually, they should be considered jointly.[5] The Jacquard apparatus's contributions to digital architectures and industrial management are inextricable. Together they establish computing's epistemological and material dependence on textile and needlecraft culture while situating this dependence in the social and economic consequences of the historical relationship between gender and labor.

Babbage recognized in Jacquard's invention the possibility of storing and executing calculations automatically, and he would later describe the functioning of such processes through the language of textile production. By the 1820s he had already invented the Difference Engine, which could mechanically execute long series of calculations through progressive operations and compile the results in printed tables. However, he saw greater value in a calculating device that could not only execute calculations based on an initial set of data and operations, but also initiate new operations during the process, with the results of ongoing calculations feeding back into the machine. He called this device the Analytical Engine. Like the Difference Engine, it would be constructed of gears, called "figure wheels." The wheels would be stacked forty-high on rows of vertical rods. Functioning on a decimal system (rather than the binary system of later, electronic computing), each wheel could be rotated like a dial into positions representing any integer from zero to nine. A numerical value with multiple digits would be rendered through the rotation of consecutive wheels on a given rod, each wheel representing a digit in the sequence. Thus the numerical value "847" would be represented by three wheels, with the first wheel rotated

A LA MEMOIRE DE J. M. JACQUARD.

**Figure 1.4**
Portrait of Joseph Marie Jacquard, entitled *A la Memoire de J. M. Jacquard*, woven by Didier, Petit et Cie by means of a Jacquard mechanism, 1839. Library of Congress photo LC-USZ62-105321.

to the eighth position, the second wheel rotated to the fourth position, and the third wheel rotated to the seventh position. Operations would be executed through studded "barrels" controlled by punched cards.[6]

The Analytical Engine was never built, in part because of the exceeding complexity of its mechanisms and the shortcomings of Victorian precision-tool manufacturing. Babbage was forced instead to content himself with simply describing the engine's properties. In addition to scores of diagrams and sketches, he relied on analogies dependent on Jacquard's apparatus and the textile industry. Babbage would claim that anyone familiar with the principles of Jacquard weaving and analytical formulas could understand his invention "without much difficulty." He identified the engine's two main components as the "mill" and the "store."[7] For any sizable textile production facility in Babbage's day, the mill was the main area of production. It required an adjacent store as a staging area for stockpiling materials throughout the production process. These goods included raw materials and thread before weaving in addition to finished textiles awaiting delivery. Similarly, the Analytical Engine's mill would be the site of execution of all operations and calculations. It would rely on its store for its materials and the stockpiling of results, whether for future operations by the mill or for delivery.

The operator of the engine need only insert material—in this case numerical values—into the store for eventual processing in the mill. In this computing machine, as in textile manufacture, when all processes were completed the finished goods—cloth or sets of numbers—would be extracted from the store. As such, Babbage conceived the store as both memory and output unit. The term "store" would remain in use through the frenzy of computing experiments of World War II. While describing the possibilities of stored-program computing in 1945, John von Neumann eschewed Babbage's terms and their industrial and cultural implications, instead describing the computer as a system of organs. In his "First Draft of a Report on the EDVAC," a founding document of modern computing, he explained that the calculating, controlling, and memory components of the computer "correspond to the *associative* neurons in the human nervous system. ... These are the *input* and *output* organs of the device."[8] Wendy Chun notes that von Neumann made this change "in order to parallel biological and computing components."[9] Von Neumann's choice of terminology attempts to naturalize the computer as a living, conscious entity capable of thought and cognition. While "memory" would become commonplace, von Neumann's evocation of the body did not readily take hold. A decade after his effort to shift computing's conceptual and semantic frame away from textile culture, computer memory itself would be manufactured as a textile of metal wires that replaced the vacuum tubes and mercury delay lines in use when von Neumann wrote his report.

As has been noted in most histories of computing, Babbage's engine would further emulate Jacquard's device in its use of punched cards as a type of read-only memory to compile and archive data that could be used to guide processing. This system for preserving and circulating data would find its greatest success, of course, in twentieth-century mainframe computing, which made the punched card synonymous with digital information technologies. In her study of the grid as a modern trope, Hannah Higgins claims that Jacquard's punched card serves as "the mechanism of transition between the soft grids of textile technology and the hardware of the information age; it translates the net from its physical expression in textiles to a modeling form that would tabulate, sort, and integrate."[10]

Textile production and the Jacquard apparatus did not contribute to Babbage's engine simply by allowing one mechanical epistemology to benefit another, however. Rather, they formed the conceptual underpinnings of that machine. The Analytical Engine required the mill and the loom to explain its processes, to make it meaningful as both concept and enterprise. This is reflected most strongly in the writings of Babbage's collaborator, Ada Lovelace, a key female presence in modern computing's early history and popularly known as the world's first programmer.[11] Perhaps Lovelace's most important contribution—more important even then the operation sequences she drafted for Babbage—was her interpretation of the Analytical Engine as a cultural object. This machine would not merely be a device for making calculations, Lovelace demonstrated; it would reframe the processes of production. The Analytical Engine was an achievement in design and interactivity, with the binary code of the punched-card system at its base. As Lovelace explains in the extensive notes accompanying her translation of Luigi Federico Menabrea's "Sketch of the Analytical Engine Invented by Charles Babbage, Esq.":

The distinctive characteristic of the Analytical Engine, and that which has rendered it possible to endow mechanism with such extensive faculties as bid fair to make this engine the executive right-hand of abstract algebra, is the introduction into it of the principle which Jacquard devised for regulating, by means of punched cards, the most complicated patterns in the fabrication of brocaded stuffs. It is in this that the distinction between the [Analytical and Difference] engines lies. Nothing of the sort exists in the Difference Engine. We may say most aptly, that the Analytical Engine *weaves algebraic patterns* just as the Jacquard-loom weaves flowers and leaves.[12]

The Analytical Engine would generate irregular patterns or forms, equivalent to the image capabilities of a Jacquard-equipped drawloom. However, Lovelace adds, "It should be remembered also that the cards, when once made out for any formula, have all the generality of algebra, and include an infinite number of particular cases."[13]

The Analytical Engine's punched cards were of three types: operation, number, and variable. Operation and variable cards had forms, but not values, allowing the production of unlimited calculations as well as the interchangeability of cards to produce new numbers. In the case of the Jacquard apparatus, a card also represented the form—a particular sequencing of rods—but not the value of an operation, such as what type and color of thread was used. Forms might be repeated several times in the production of a particular textile motif. By making selections from existing sets of cards for the Jacquard apparatus and arranging the selections in different combinations, weavers potentially could vary the vertical sequence of forms in a textile to produce any number of designs. This occurred in the production of bed coverlets in mid-nineteenth-century America, for example. American weavers attached Jacquard apparatuses to handlooms and bought mass-produced mix-and-match card sets, like consumer-grade software programs, to produce their own, unique patterns.[14]

Indeed, a Jacquard-equipped loom functioned much like a programmed computer. The machine could be adapted to quickly perform any number of calculations simply by running punched-card programs. It could not, however, store any calculations or operations during the production process. That is, the system had no memory. If a calculation or operation needed to be repeated in the mechanism, it would have to be repeated in the chain of cards, substantially adding to the chain's length. Any deviation in the results from one passage of the cards to the next—from one woven piece to another—would be the product of material flaws (e.g., rod-card misalignment or thread inconsistencies and breakage) or human error (e.g., skipping a passage of the shuttle), rather than any change in calculations during the operation. Lovelace saw a symbiosis between loom and engine that would bring them closer together in their conception and functioning, however. In 1834, the year Babbage first envisioned the Analytical Engine, Lovelace visited English textile mills with her mother, Anne Isabella Byron, where she saw firsthand how punched-card systems contributed to silk ribbon production.[15] If the Analytical Engine could retain data obtained from punched cards for later access and use, Lovelace believed a similar system could be designed for the loom, to eliminate the necessity of repeating the same commands or patterns multiple times within any chain of cards. In speculation that extends beyond Babbage's own interest in the Jacquard apparatus and its links to computing, Lovelace explained that this weaving device could incorporate a further component, allowing for the storage and reintroduction of specific cards in the train of cards during production of a textile. She states:

> It has been proposed to use [backing] for the reciprocal benefit of that art, which, while it has itself no apparent connexion with the domains of abstract science, has yet proved so valuable to the latter, in suggesting the principles which, in their new and singular field of application, seem likely to place *algebraical* combinations not less completely within the province of mechanism, than are all those varied intricacies of which *intersecting threads* are susceptible. By the introduction of the system of *backing* into the Jacquard-loom itself, patterns which should possess symmetry, and follow regular laws of any extent, might be woven by means of comparatively few cards.[16]

Lovelace's description synthesizes the logic behind the processes of weaving and assembly. In form and value the Analytical Engine's functioning bears similarities—at least at the level of systems logic—to block-pattern textile assembly, for example. A block pattern is a basic template upon which any number of other garment patterns can be produced. The block pattern can be used to produce a specific object from multiple pieces of fabric or serve in the production of a variety of objects. In the Analytical Engine, cards could be "backed" in groups or batches to be used multiple times within a single operation or set of calculations. Natalie Rothstein notes the same with the Jacquard apparatus, where "the pattern could be changed in a few minutes, provided the cards were cut and laced together."[17] A set of cards used within an operation could therefore be reintroduced at a later point within the process to produce a pattern, while nevertheless modifying results, just as a block or motif pattern could be used repeatedly within production to create identical or different results,

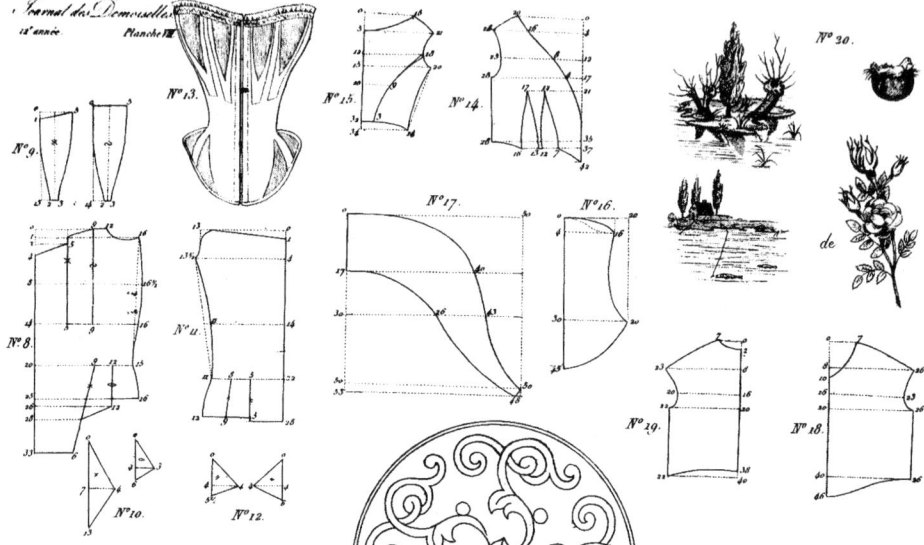

**Figure 1.5**
Pattern for a bodice, published in *Journal des Demoiselles*, August 1844.

depending on how this template was used in relation to other patterns. Just as the Analytical Engine's card system represents the basic formula or program upon which variations and synthesis can be introduced by entering different data, block patterns represent the material base upon which variations and synthesis of woven material can be produced to create garments.

Commercial patterns for home sewing began appearing in British and French periodicals during the same decade in which Babbage conceived the Analytical Engine (figure 1.5). Patternmaking manuals and mass-produced patterns would become a regular part of home-based textile and needlecraft culture in nineteenth-century Europe and North America, standardizing domestic production while allowing women to exchange patterns or work collectively on garments, even when separated by great distances.[18] In other words, patternmaking functioned as the home industry equivalent of punched cards by allowing reproducibility of results while serving as the base for variation. This was aided greatly by the spread of the sewing machine as a household appliance in the second half of the nineteenth century. Originally meant for use by men in large-scale manufacturing, the sewing machine became one of the century's greatest home consumer successes in a process of technological diffusion that would be matched a century later by the computer's shift from the mainframe and business software to home electronics and consumer software. As Tim Putnam notes, "Sewing machines had been designed for the workshop and only became home appliances when claimed by women as their own."[19] Together, the sewing machine and patternmaking brought textile culture into the home in new forms as mass-produced cloth was assembled at home either as piecework labor in the garment industry or as a leisure activity encouraged by the fashion plates and patterns found in women's magazines.[20]

Only a few decades earlier, automated production on machines equipped with mechanisms similar to Jacquard's device had brought textile production *out* of the home and workshop and into the factory, becoming a well-known sector of the Industrial Revolution. Histories of computing rarely broach the impact of Jacquard's invention on mass production and gender differentiation in the industrial workforce, yet it had lasting consequences that would shape the nature of the electronics industry—and its dependency on the textile industry—over a century later.[21] While women had woven at home for thousands of years, commercial textile production—often of larger and more elaborate pieces than could be reasonably made on a domestic loom—had been the domain of men. Pre-Jacquard silk weaving was considered a skilled profession, requiring careful calibration of the loom and calculation of warp configurations for each pass of the shuttle.

The Jacquard apparatus is emblematic of the generalized upheaval in production methods and labor requirements created by increased standardization, mechanization, and automation in the nineteenth century. When Jacquard introduced his device into the French silk industry in Lyon, weavers sabotaged it and made threats against his life.[22] It first came into use in Britain in English cotton mills around 1813 but took a decade to definitively penetrate the British textile industry, perhaps spurred by imports of cheaper, Jacquard-made French silks during the mid-1820s.[23] Nevertheless, mechanized power looms had begun to replace handloom weaving in Britain by the early years of the nineteenth century. The Luddite movement of the 1810s, protesting working conditions in the weaving and knitting industries, led to waves of frame smashing in northern England, including steam-powered looms that had been newly installed in Manchester mills.[24] Parliament's House of Lords, which included Lovelace's father, the poet George Gordon Byron, claimed that such revolts were the result of "the use of a new machine, which enabled the manufacturers to employ women, in work in which men had before been employed."[25] Though this was an oversimplification, the power loom had allowed commercial weaving to be recategorized as semiskilled or unskilled work, a change that facilitated the employment of women.

None other than Babbage would argue the benefits of this development in his widely read *On the Economy of Machinery and Manufactures*, published only two years before he conceived of the Analytical Engine as a computational transfiguration of the textile mill. Focusing on textile industries, Babbage explained that increased mechanization "does not ... invariably throw human labour out of employment" but rather "enables children and inferior workmen to execute work that previously required greater skill."[26] Noting that "during the whole of this period the wages and employment of handloom weavers have been very precarious," he reasoned that "a diversity of employment amongst members of one family will tend, in some measure, to mitigate the privations which arise from fluctuation in the value of labour."[27] The mechanization of textile production opened a door in this regard: "A hand-weaver must possess bodily strength, which is not essential for a person attending a power-loom," Babbage remarked. "Consequently, women and young persons of both sexes, from fifteen to seventeen years of age, find employment in power-loom factories."[28]

In the semiautomation of the loom and subsequent changes in required skills, the principles of computing and gender-defined labor structures entered into what would become a looped pattern. Women, especially those who were young and rural, would constitute the favored labor pool for nineteenth-century textile industries in Europe and North America. The pattern would repeat in the electronics industries of the twentieth and twenty-first centuries, where young, rural women have been regularly sought for assembly work, particularly in Asia.[29] By the late 1850s, for example, 83 percent of workers at the Courtauld silk mill in Halstead, East Anglia, were women. Not a single man out of the mill's more than one thousand employees tended one of its power looms. The few male weavers living in Halstead were employed instead in soft silk handloom weaving workshops.[30] Women textile workers in Halstead and elsewhere were paid less than men due to their "unskilled" status—winding and weaving for 15–50 percent less than what the mill's male clerks, overseers, and mechanics earned—even as the industry benefited from women's informal home training in needlework, sewing, weaving, and other textile-related crafts. As such acquired skills were considered part and parcel of homemaking and women's work, there was no attempt to acknowledge or reward them through a worker's status and compensation.[31] In a well-known American example, industrialist Francis Cabot Lowell exploited similar socioeconomic conditions in mid-nineteenth-century Lowell, Massachusetts, by recruiting young New England farm women to work in his urban textile mills, replacing home and hearth with dormitory and dining hall. The Lowell "mill girl" became a symbol of American industrial ingenuity. While men in the American textile industry negotiated their wages, mill girls were paid at fixed rates "high enough to induce women to leave the farms ... but low enough to offer the owners an advantage in employing women rather than men." This led to collective actions by the Lowell mill workers in the 1830s and 1840s in protest of meager wages and deleterious industrial conditions.[32]

## Weaving Core

The "unskilled" domestic training of women in textile and craft practices returns repeatedly in the culture and economy of modern computing. Just as the processing and memory coupling of Babbage's mill and store has remained the basic paradigm of computing structures, textile-based manufacturing practices have continued to be the organizational paradigm behind digital production. Hand sewing and weaving were key design components of two of the most significant computing projects of the twentieth century—the invention of a real-time, interactive computer at MIT in the 1950s, and the construction of the navigational systems that took men to the moon and back in the 1960s. Both of these projects, to be considered here, are prominent historical examples closely related to broader practices of electronics assembly based on textile assembly techniques.

Recalling that neither Babbage's Analytical Engine nor any comparable device was built in the nineteenth century, for much of the first half of the twentieth century—and especially during World War II—but in some cases even into the 1960s, computers were the minds and pencils of women executing calculations in government bureaus, university laboratories, and similar research settings. In 1945, for example, nearly two hundred women were employed in this capacity during the construction of the U.S. Army's

Electronic Numerical Integrator and Computer (ENIAC).[33] After the end of World War II, the computer room at the United States Bureau of Standards, where the Standards Western Automatic Computer (SWAC) would be developed, was still filled with desks and chairs.[34] "On each desk was an electrically powered mechanical calculator operated by a skilled woman," recalls David Rutland. "Each woman had a work sheet with the numbers that she was to use in her calculation in the left-hand column. ... Across the top of the other columns were listed the operations. ... The results of some operations became the input data for the next operation."[35] Each woman's sheet was not only a grid of information, but one that represented a logical pattern of operations. As a system for recording data during operations, these sheets functioned as a form of memory, the store to the mills of the mind and mechanical calculator.

Jennifer Light has demonstrated how the transition from human computers to electronic machines at the end of World War II was initially accompanied by the creation of another feminized computing occupation: the computer "operator."[36] Female computers or clerical staff familiar with business machines became operators of mechanized systems such as ENIAC, until they were replaced by men after the war and the position was renamed "programmer" to mark a shift in both gender and status. Operators, like the first programmers who followed them, were required to understand and troubleshoot software as well as hardware. As operator Betty Jean Jennings explains, "Since we knew both the application and the machine, we learned to diagnose troubles as well as, if not better than, the engineer."[37] Among the greatest hardware problems was memory. Mercury delay lines, vacuum tubes, and other forms of storing data during this period proved to be inefficient and unreliable, requiring frequent maintenance and jeopardizing the effectiveness and accuracy of programs and calculations.

World War II had brought major breakthroughs in analog and digital computing as governments devised algorithmically functioning machines for making and breaking codes, predicting ballistics trajectories, and executing other tasks tied to the complexities of military operations. This growth only accelerated with the onset of the Cold War.[38] The need for reliable random-access memory within computing systems to support faster calculations, greater capacity, and flexibility in operations, would eventually fold women's textile labor back into the process of technological development. The "single most important computer project of the postwar decade," according to computer historian Paul Edwards, was the Whirlwind computer developed at MIT's Lincoln Laboratory.[39] Originally organized to produce an analog flight simulator, the project grew into the digital computer system behind the U.S. Air Force's Semi-Automatic Ground Environment (SAGE) air defense systems. Whirlwind led not only to real-time graphical screen interfaces, handheld optical input devices, and other elements that would become common to computing later in the century, but also to a new form of memory. Coincident current (or static) magnetic matrix storage, known as magnetic-core memory, was based on principles of electric current and magnetization. It quickly became the leading form of computer memory and remained standard into the 1970s.[40]

Core memory's material form was a wooden or metal frame strung with a taut grid of fine wires (figure 1.6). A small ferrite ceramic ring or "core" was suspended at each intersection of these wires (figure 1.7). This pattern of construction incorporated three types of wires: driving, sensing, and inhibiting. Driving wires would form the horizontal and vertical lines of the grid that held each core in place, while the sensing wire would be threaded diagonally at their intersections. The inhibiting wire would be threaded back and forth horizontally through each row of cores. Thus four wires passed through each core. Applying current through driving and sensing wires would produce a clockwise or counterclockwise charge to each core, depending on the direction of the current passing through it. Rings could be magnetized as positive or negative, equivalent to a one or zero when read by the computer's processor. The sensing wire allowed the binary data to be read, while the inhibiting wire prevented changes in polarity where necessary. Unlike other memory capacities developed at the time, a core's charge—that is, its memory—would remain even when the computer's current was disrupted or cut.[41] Several groups and institutions had been developing core memory independently, but Lincoln Laboratory was the most successful, building a prototype in 1952 and converting the Whirlwind's memory system from electrostatic cathode ray tubes to core by mid-1953.[42] "In five years' time, core memory would replace every other type of computer memory," explain Martin Campbell-Kelly and William Aspray. "The value to the nation of the core-memory spin-off alone could be said to have justified the cost of the entire Whirlwind project."[43]

Project director Jay Forrester sketched the basic configuration for core memory in 1949 and left it to MIT graduate student William Papian to

**Figure 1.6**
Core memory plane from Project Whirlwind, Lincoln Laboratory, MIT, c. 1953.
Courtesy of the Computer History Museum, CHM image 102622505.

design a functioning system.[44] A single, $16 \times 16$ prototype array was built
in 1952 and thoroughly tested on a specially built computer called the
Memory Test Computer.[45] By early 1953 the prototype had been deemed
a success and core memory was ready to be implemented, first in $32 \times 32$
arrays, then in larger, higher-capacity $64 \times 64$ arrays.[46] Stating in the project's biweekly report of January 2, 1953 that "design of the memory planes
was completed, and construction of the mounting frames is in progress,"
the Memory Section hired Hilda G. Carpenter as a laboratory assistant and
technician responsible for assembling the intricately patterned frames of

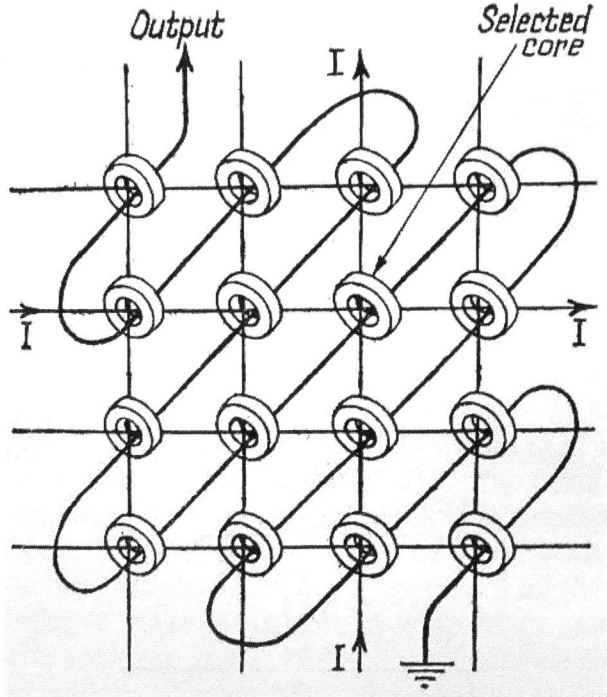

**Figure 1.7**
Diagram of core memory threading. From M. V. Wilkes, *Automatic Digital Computers* (London: Methuen, 1956).

wires and rings.[47] A report on memory plane construction from the following summer asserted that seventeen memory planes had been produced. A photograph appended to this report shows two women threading and testing planes at a lab bench (figure 1.8). Even with the aid of an assembly jig, a plane required four days to wire.[48] Each plane was composed of 1,024 cores suspended diagonally and turned in alternate directions in the 64 × 64 grid of wires.

"The final design was both simple and elegant," remarked British computer pioneer Maurice Wilkes, who called core memory "a brilliant achievement."[49] Bernard Widrow, an MIT graduate student involved in the implementation of Whirlwind core memory planes, would explain decades later: "This all had to be hand-wired. All the wiring in this memory plane was done by a woman who was a technician working in the lab. I don't

**Figure 1.8**
Project Whirlwind staff threading and testing core memory planes, Lincoln Labo-
ratory, MIT, 1953. Photograph used and reprinted with permission of The MITRE
Corporation. © 2016. All other rights reserved.

remember her last name. But her first name was Hilda. And Hilda wound
all these memory planes. It's like knitting. ... Hilda wove all those wires.
It's like weaving."[50] While Widrow's comments demonstrate esteem for the
skill involved, his recollection of a first name, but not a last, reflects Carpenter's relatively low status as a laboratory assistant. Except to record her
hiring, her name does not appear in the reports and memoranda of Project Whirlwind. She does appear as a model in an article on core memory
production in a 1956 issue of the journal *Electronics*, however. Carpenter,
a woman of color, is seen pouring cores into a sorting tray and weaving a
plane of cores with a hypodermic needle. Yet even in this article, her name
is absent.[51]

**Figure 1.9**
Detail from figure 1.8. Project Whirlwind staff member threading a core memory plane, Lincoln Laboratory, MIT, 1953. Photograph used and reprinted with permission of The MITRE Corporation. © 2016. All other rights reserved.

A memory plane would require about forty hours to fully wire in 1953 (figure 1.9).[52] The design and production of core memory drew on properties and practices of weaving, needlepoint, embroidery, and beading (figure 1.10). Cores were mounted like beads by "stringing" them on a driving wire. The grid of driving and inhibiting wires followed the form of traditional weaving, although technically they were not woven since the weft did not follow an over-under alternation with the warp but was merely laid over it. This technique was common to twentieth-century craft weaving, however. Craft weaving was popularized through mass-produced handloom kits, including Easiweave, Weave-It, Magic Loom, and the Lily weaving loom (figure 1.11). These kits typically contained a wooden, metal, or plastic six-inch square pinframe remarkably similar to the frame of the core plane. "Easiweaving is a novel and modern combination of two of the oldest handcrafts—weaving and needlework," one guidebook explained.[53] Weaving on such frames was a four-step process that began with threading two layers of yarn, first horizontally, then vertically (figure 1.12). The core memory plane's grid of driving and inhibiting wires resembled this pattern. The plane's added suspension of the core rings at the intersections of wire followed techniques of on-loom bead weaving, which locks a bead

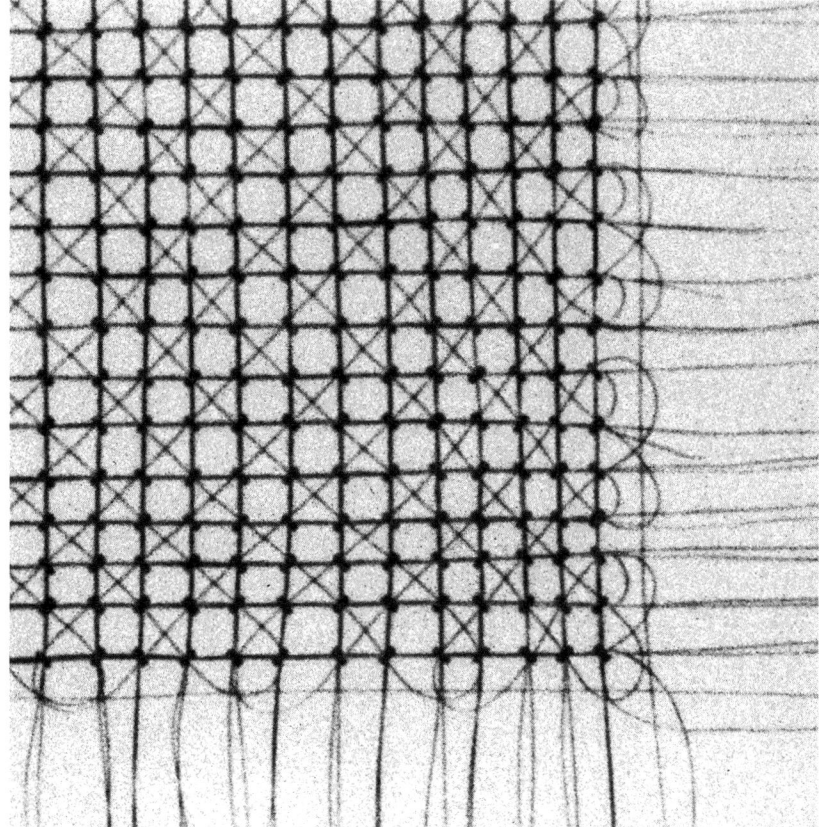

**Figure 1.10**
Detail from figure 1.6. Portion of a core memory plane from Project Whirlwind, Lincoln Laboratory, MIT, c. 1953. Courtesy of the Computer History Museum, CHM Image#: 102622505.

within the grid of warp and weft threads. The multiple threading of driving, sensing, and inhibiting wires accomplished this result. In fact, bead work instructions in twentieth-century hobby books suggest that weaving beads on a handloom be done only with "lengths of heavy linen thread or fine wire" capable of supporting the weight of the beads.[54] The diagonal threading of the sensing wires through the cores of the memory plane, for its part, has its roots in needlework, which regularly employs diagonal stitching across an underlying straight- and cross-grain weave. This detail draws on the logic of the final two steps in craft kit weaving, which rely on

**Figure 1.11**
Cover of *Lily Weaving Loom Instruction Sheet*, c. 1955.

needlework techniques of stitching through the threaded base to produce the weave of the textile. A similar bias stitch, diagonal weave orientation was available in some handloom kits, including the Bias Weave-It.[55] In both home weaving and core weaving, once the block was completed, it could be removed from the frame as a loose piece of fabric, to be sewn into patterns with other blocks or, in the case of memory, stretched onto a permanent frame to be soldered and mounted in the computer.[56]

When Lincoln Laboratory contracted IBM in 1953 to design the Whirlwind II—soon inelegantly renamed the AN/FSQ-7—the company embarked

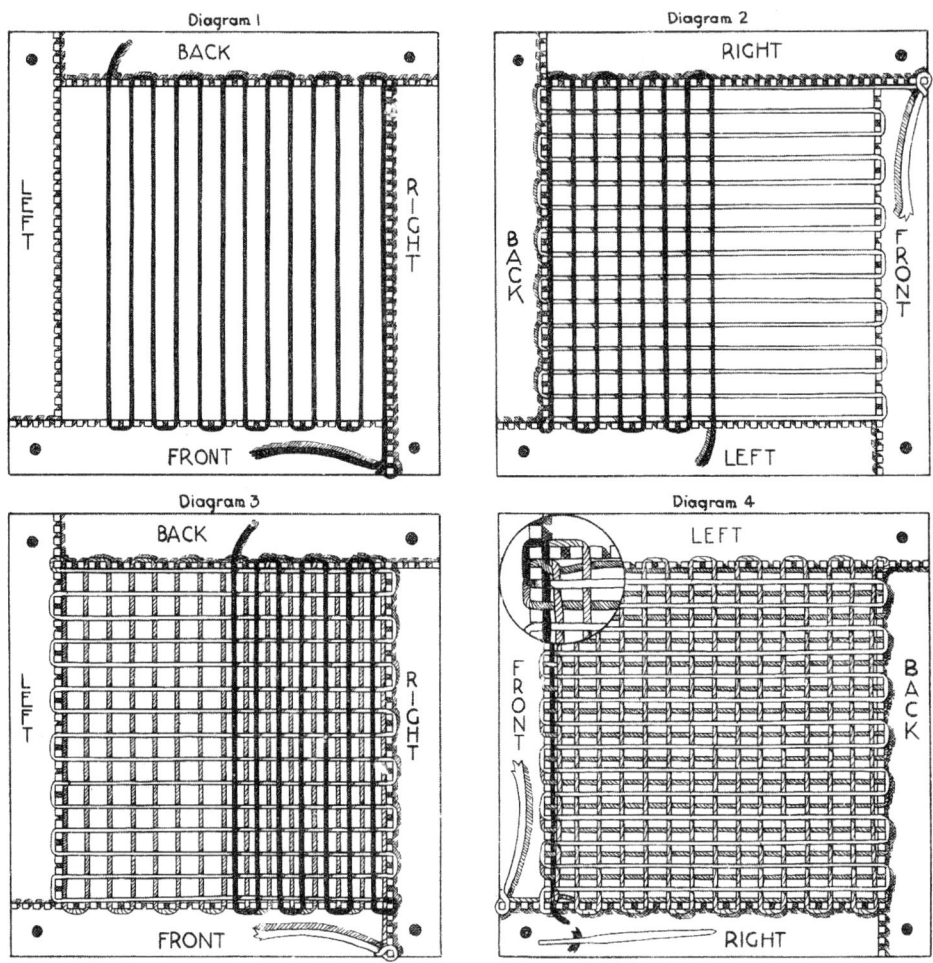

**Figure 1.12**
Diagram of threading and weaving processes on the Cynthia Easiweave Frame, c. 1935. From *Cynthia Easiweave Frame Directions*.

on a "crash research program" to improve core memory's reliability and speed. Core memory's superior properties when compared with other memory systems would contribute to the sales success of the IBM 704 in the mid-1950s.[57] As Michael Williams claims, "The commercial availability of magnetic core memory was the great watershed point in the development of computers."[58] Core memory required less space, less maintenance, and thus less downtime than other systems. The first successful minicomputer, the refrigerator-sized PDP-8 released by Digital Equipment Corporation in 1965, relied on transistors and core planes to minimize size while maintaining capacity.[59] Core memory would remain common into the early 1970s, when IBM finally halted development to concentrate on semiconductor technology.

Despite its success, core memory production remained slow, intricate, and laborious when compared with other computer components. To simplify and increase core memory production, IBM spent several years devising specifications for machine weaving. Fully mechanized production was difficult, however, given the configuration of materials and coordination of steps (particularly the bias stitch of the sensing wire), and a viable method would not be available until after core memory had been definitively overtaken by semiconductors.[60] Attempts to produce machine-woven, "screen memory" grids not requiring core beading also met with little success (figure 1.13).[61] Ultimately, IBM decided that its core matrix manufacturing specifications could be adequately applied and achieved through manual production. Rather than automated production machinery, IBM opted for a number of discrete instruments that followed the logic of home craft kits to simplify hand-wiring and eliminate errors in threading and configuring the core "beads." Described at length in the *Electronics* article for which Carpenter posed, this would remain the principal means of making core memory.[62] In this process, a plane weaver would pour loose cores into a matrix tray and manually sift them into the pattern mold to secure their proper orientation. She would then hand-thread the cores with wire inserted into a needle feeder at one edge of the frame. Once a line of cores was threaded, the wire would be taken up by the clamps of the wire wrapper fixed at the opposite edge of the frame. Molded plastic frames with grooves along the upper edge—almost identical to the plastic frames in Easiweave and other home weaving kits—would hold the wires.[63] This process, known as "winding" core, was little removed from either Carpenter's original work or

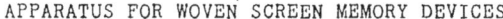

APPARATUS FOR WOVEN SCREEN MEMORY DEVICES

Filed Nov. 12, 1963                                    6 Sheets–Sheet 1

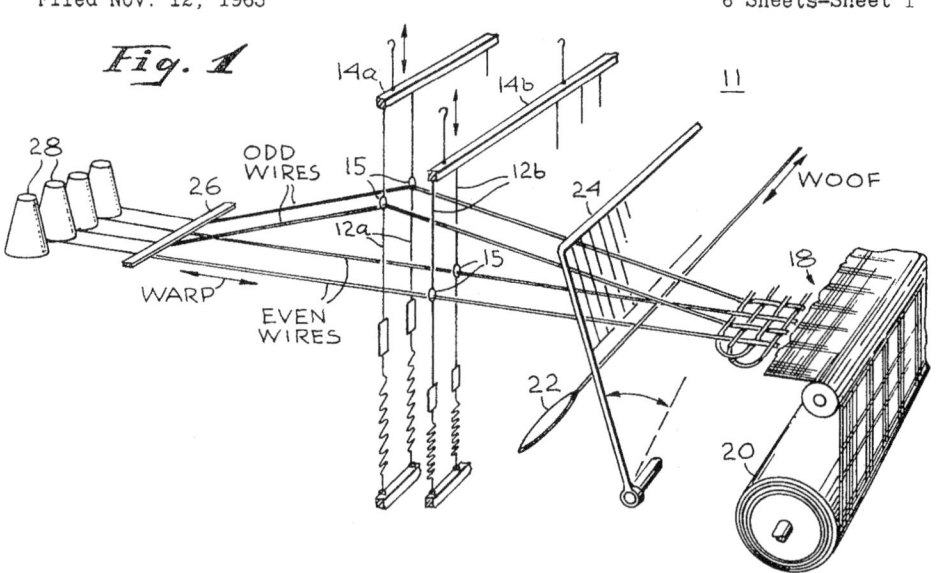

**Figure 1.13**
Diagram of a loom for weaving screen memory. From D. R. Boles et al., Apparatus for
Woven Screen Memory Devices, U.S. Patent 3377581 A, 1968.

popular handloom techniques. Further attempts to simplify and improve
the material aspects of core memory production concentrated on changing
the weaving pattern itself. A new pattern developed by IBM in 1962 allowed
one wire to serve as both a sensing and inhibiting line, reducing from four
to three the number of wires threaded through each core. In the new pattern, the sensing wire shifted across two rows as it passed the central axis of
the plane. This alteration had emerged during attempts to thread sensing
wires in figure eight patterns requiring a shift by one row.[64]

Because the labor-intensive aspect of core memory production raised
costs, IBM included steep markups on memory, which could be rented or
purchased, in its larger accounts.[65] The price of core memory hinged on the
cost of weaving, as the cost of manufacturing cores dropped from thirty-
three cents per core in 1953 to a fraction of a cent by the early 1960s.[66] To
maximize profit margins, IBM relocated core memory production to Japan
and Taiwan in 1965, foreshadowing the practice of outsourced electronics

manufacturing that would become routine for American companies later in the century.[67] The description in one history of IBM of how these events unfolded reads like a neoliberal colonialist fantasy:

> Taking bags of cores, rolls of wire, and core frames to Japan, [an IBM plant manager] returned ten days later with hand-wired core planes as good as those that had been wired by the automatic wire feeders in the Kingston [New York] plant. … It was slow, tedious, meticulous work to string wires through each of the thousands of tiny cores in each core plane, but the cost of labor in the Orient was so low that production costs were actually less than with full automation in Kingston.[68]

Much of a decade's work in automating multiple aspects of the weaving process would be abandoned. Core weaving for IBM and many of its competitors was thereafter achieved through Asian manual labor.

Alongside the production of core memory for the world's mainframe computers (most of which were IBM systems), one of the most prestigious and expensive postwar computing projects was the development of hardware and software for the U.S. government's Apollo moon landing program. MIT's Instrumentation Laboratory and Raytheon Corporation collaborated on the digital guidance and navigation systems required to send humans to the moon. Here again, weaving and women's labor were critical to the success of the project, in this case through a little-used variation of core memory production. Navigational computing on board the command capsule and lunar module relied on standard core memory planes for its erasable memory, but the fixed memory guidance programs were constructed through a much rarer "rope" form of core weaving. Rope memory offered a secure and durable yet flexible and compact construction that reduced space requirements within the spacecraft's fuselage while increasing the memory's resistance to vibration, shock, and other physical risks associated with space travel.[69] Unlike the stacked-plane configurations of mainframe core memory, rope memory consisted of strands of cores threaded with relatively loose lengths of bunched wires (figure 1.14). Upon this base, sensing wires were threaded either through a core to produce a positive bit (one) or around a core to produce a negative bit (zero).[70] Contrary to woven memory planes, rope memory prohibited a change in a core's charge, since the difference between positive and negative charges would derive from the configuration of the threading itself. "This fixed memory is actually composed of magnetic cores with wires woven in and out, sewn in with a pattern, where the information … is in the pattern of the sewing," Instrumentation

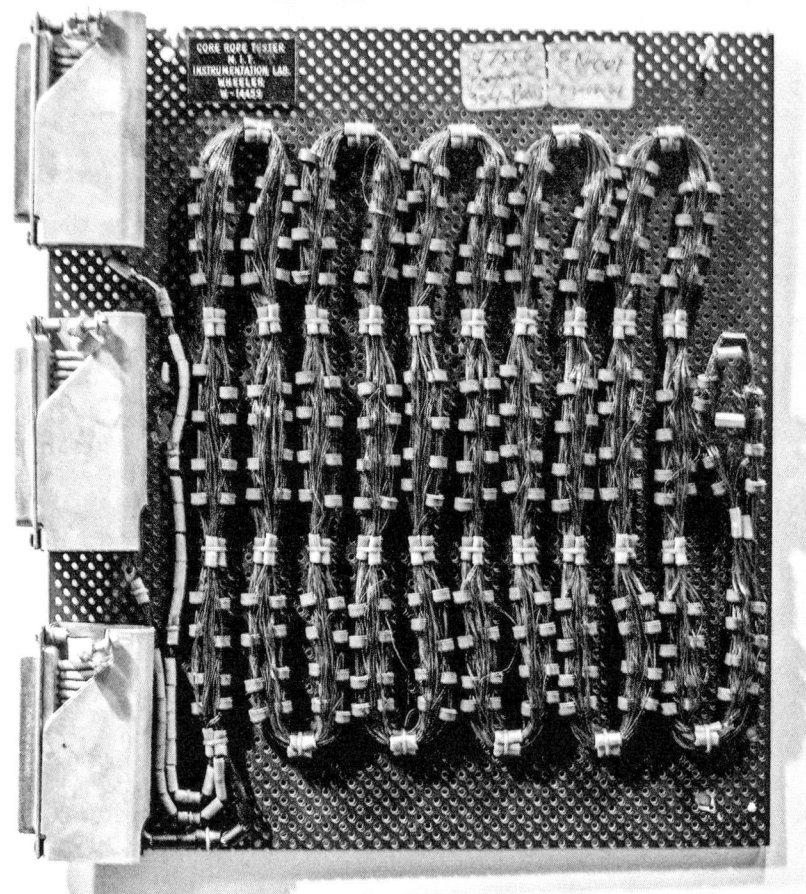

**Figure 1.14**
Apollo guidance system rope memory, mounted on a test panel, Instrumentation Laboratory, MIT, 1961. Photo by Nova13.

Lab Assistant Director Albert Hopkins evocatively explained.[71] Up to sixty-four sensing wires could be threaded through and around a single core, in any sequence, creating a far denser storage system than core memory planes, with up to four sets of 16-bit words per core.[72]

"We have to build, essentially, a weaving machine," systems designer Ralph Ragan told reporters as the construction of the guidance systems got under way. Raytheon hired retired or laid-off women textile workers to execute the task in its Waltham, Massachusetts, plant. The women represented

New England's dying textile industry, which had flourished in Lowell and surrounding towns and cities but would disappear by the 1970s in the face of competition from Japan and elsewhere.[73] These memory weavers worked alone or in pairs with the gridlike "weaving machine" that was little more than a frame holding the rope in place while they threaded the driving wire back and forth through it, like a weft shuttle, to weave the code into the cores.[74] While essentially beading, this handiwork also followed principles of embroidery already found in the threading of the sensing wire in grid-based core memory production. In both cases, new patterns of thread were added to an existing, invariable configuration. "We called it the LOL method. The 'little old lady' method of wiring these cores," states Richard Battin, an MIT Instrumentation Lab director involved in the project.[75] Twenty miles down the road from Lowell, the nineteenth-century mill girl had been replaced by her aged counterpart. As David Mindell explains, "NASA was well aware that the success of its flights depended on the fine, accurate motions of these women's fingers."[76]

## Stitching Boards

While core formed the base of Apollo's guidance system memory, the guidance computer's processor was comprised of four thousand to five thousand of the first silicon chip integrated circuits for digital computing. By the time the Apollo program ended in the mid-1970s, silicon chip capacitor memories had replaced core as the preferred form of random-access memory. Unlike core, a silicon chip semiconductor's microscopic circuit patterns are not threaded by hand but imprinted through photolithography, layer by layer, into the wafer. To complete the circuit, minute quantities of aluminum or another conductive metal are poured into the imprinted channels.[77] Full circuits on small chips were fundamental to the development of the personal computer and remain integral to nearly all mobile, digital devices. Even with the automated mass production of silicon chips, however, women have remained primarily responsible for their assembly into larger electronic components through the soldering of integrated circuit boards and microchips. The relatively short history of core memory production would serve as an influential precedent in this change, having established the place of "unskilled" women's labor and the role of the global textile industry in the assembly of components basic to a computer's functioning. Assumptions concerning women workers' "natural" ability with small parts and meticulous, repetitive work would rest on the historical examples of electronic and textile production from the nineteenth and twentieth centuries, as well as on the unacknowledged and informal training of the home.

With the integrated circuit, the gendered and racial divisions already exploited by core production became more evident, expanding on a global scale. In the 1960s and 1970s, Fairchild Semiconductor—the "mother firm" in California's Silicon Valley, the hub of integrated circuit innovation— sought to exploit the weaving skills and limited employment opportunities of Navajo women on reservations in the U.S. Southwest by building an assembly plant in Shiprock, New Mexico.[78] Fairchild's promotional materials for the ill-fated Shiprock plant drew visual correspondences between the aesthetics of Navajo rugs and the design of integrated circuits.[79] Exploring the example of Shiprock in relation to the racialization of electronics manufacturing, Lisa Nakamura points out that Fairchild management believed Navajo weaving traditions gave women on the reservation the "natural" capability to "visualize complicated patterns and ... memorize complex integrated circuit designs."[80] Fairchild's promotional materials pointed out: "A Navajo woman weaves a perfectly patterned rug without ever seeing the whole design until the rug is completed. ... The blending of innate Navajo skill and Semiconductor's precision assembly techniques has made the Shiprock plant one of Fairchild's best facilities—not just in terms of production but in quality as well."[81] A journalist covering the beginnings of the plant explained that "the same adroit fingers which have made the Navajo Indians famous for fine rugs and jewelry are now turning out one of the indispensable adjuncts of this electronic age."[82]

Supported by federal government wage subsidies, the Shiprock plant employed mostly women in soldering chips and bonding wires. It assembled some of the early semiconductors used in the space program as components for the Apollo Saturn V rocket. After layoffs in the mid-1970s, however, the pall of colonialist exploitation by a federal government in collusion with a major corporate force in the military–industrial complex led to a well-publicized takeover of the plant by the American Indian Movement.[83] Though the event lasted only a few days, Fairchild closed the plant shortly after. By that time Silicon Valley had become the flourishing center of semiconductor research and manufacturing, with dozens of companies spinning out of Fairchild.[84] Much like Shiprock, however, 70–75 percent of semiconductor assemblers in Silicon Valley were women. Forty percent of these women were minorities.[85] In the 1980s, semiconductors produced in Silicon Valley were shipped to Asia to be bonded to circuit boards.[86] Production of semiconductors and circuit boards, like core memory production before it, migrated to Asia by the 1990s in what some American industry analysts called "the commodity memory debacle."[87]

In truth, Fairchild and its domestic competitors had moved overseas much earlier, just as IBM had, by establishing plants in Hong Kong in 1963, Taiwan in 1964, South Korea in 1966, Singapore in 1968, Malaysia in 1973, and the Philippines in 1974.[88] Whether owned by multinationals or local industrialists, such electronics facilities in Southeast Asia and China continue to rely heavily on women's manual labor to assemble integrated circuits.[89] In his book on the rise of the Japanese consumer electronics industry, Simon Partner establishes the historical affinities between textile and electronics production in Asia, arguing that the success of the electronics industry in Japan ultimately rested more on its relation to the country's textile industry than its technological innovations. According to Partner, success "derived not from the technological content of the products produced—that remained largely irrelevant—but from the availability of cheap labor to make products that did not easily lend themselves to automated manufacturing techniques."[90] In particular, Japanese electronics plants hired the same young, female workers regularly recruited out of middle school by textile mills. When IBM outsourced core memory production to Japan in the mid-1960s, the country's textile industry—supported mainly by this workforce—already had an international reputation for "social dumping," that is, producing and exporting woven goods at prices that suggested unlivable wages. "The similarity between the female workers in transistor and radio factories [of the 1960s] and those who had traditionally constituted the foundation of the textile industry is striking," Partner asserts. "Underneath the apparent revolution in technology development and industrial structure lay a profound continuity based on the abundance of extremely cheap, relatively docile female labor."[91] The young women in these electronics plants—often teenagers—would sit at assembly tables for hours, putting together components with tweezers for extremely low pay when compared with the earnings of workers in male-dominated industries. They were known as "transistor girls" by company management, a term that reemerges in twenty-first century discourse around gender, race, and electronics assembly through the figure of the Chinese "iPhone Girl."[92]

Gendering meticulous handwork and patience as female is an ideological construct extending across industries, cultures, and regions. It has thrived in the textile and electronics industries in Europe, Africa, the Americas, and Asia. An American management manual from the 1970s makes some general observations: "Many women are well suited for using precision tools, inspecting products, typing, and assembling small or intricate parts. ...They have more patience and adjust better to routine work—and they will stay with it longer than men. They are willing to give more attention to small details and exacting work."[93] Plenty of examples from the sociological and anthropological work done on gender and electronics labor around the world from the late twentieth century to today confirm such attitudes. A woman working at a Silicon Valley semiconductor assembly plant during the 1970s similarly observed: "Women make the best workers at this kind of thing because you have to be patient, you have to be good with your hands and the work is so tedious. Isn't raising children and doing housework tedious? I mean, women are good at this."[94] In a study of Malaysian electronics workers in the late 1980s, Les Levidow found that the "unskilled" labor of component assembly was done by women despite high unemployment among local men, in part because the work was feminized by manufacturers who found women to be "naturally better suited to the routinized work of the electronics assembly line: nimble fingers, acute eyesight, greater patience."[95] In *The Integrated Circus: The New Right and the Restructuring of Global Markets*, Patricia Marchak notes that studies of the success of electronics production and assembly in Asia have found that "patriarchal cultures and non-industrial lifestyles [there] ... train young women to become manually dexterous, a trait frequently cited by employers as a reason for preferring women over men in textile and electronics work."[96] In Asia today, young women are often recruited from rural areas to work in production facilities known as "assembly houses," a term underlining the domestic inflection of such work. In her firsthand experience as a worker in an electronics plant in Shenzhen, China, sociologist Pun Ngai estimates that 90 percent of the plant's assemblers were women, while an equally high percentage of the management were men. Reflecting the attitudes expressed earlier by American management specialists, she explains: "As usual, assembling tiny electronic components was often considered women's work because it required patience, care, sharp eyes, and nimble fingers."[97] In "'Nimble Fingers Make Cheap Workers': An Analysis of Women's Employment in Third World Export Manufacturing," Diane Elson and Ruth Pearson explain that the skill behind such production, because it is often acquired informally at home through lessons in sewing and needlework, is itself "socially invisible and privatized," allowing it to be construed in public and corporate spheres as "attributable to nature, and the jobs that make use of it are classified as 'unskilled' or 'semi-skilled.'"[98]

Despite the prevalence of surface-mount technology and wave soldering that embeds elements in circuit boards today through automated machinery, most contemporary electronics still require some hand assembly. Though sites of production may have changed, the example of the intersection of textile work with high-tech and appliance manufacturing given in a 1976 article on international subcontracting still resonates in today's electronics industries. "In Morocco, in six weeks, girls (who may not be literate) are taught the assembly under magnification of memory planes for computers," Michael Sharpston notes. "This is virtually darning with copper wire, and sewing is a traditional Moroccan skill. In the electrical field the equivalent of sewing is putting together wiring harnesses; and in metal-working, one finds parallels in some forms of soldering and welding."[99] Like the process of sewing together precut pattern pieces of fabric, workers in Asian assembly houses today typically perform such tasks as hand-feeding wire leads into the etched channels of printed circuit boards, soldering them, and snipping away excess wire. They align these boards with integrated circuit bonding machines that install the microcontrollers needed to ensure the device's proper functioning. A microcontroller "is 'stitched' to the board in a process that looks not unlike sewing with a sewing machine," according to one manufacturer's account.[100]

## Gestures of Memory

The conditions of assembly houses occasionally have made headlines, as with the viral coverage of the iPhone Girl in 2008.[101] In this case, a test photo taken of an unidentified, smiling woman assembler was inadvertently left in the memory of a new iPhone as it left a Chinese production facility. Finding the image after he purchased the device in Britain two weeks later, the iPhone's owner posted the image online, where it quickly

drew worldwide attention.[102] In a detailed study of the incident, Seth Perlow successfully illustrates the web of disjunctions and correspondences
that typically exist between assemblers and users through the materiality
of such devices and the different forms of labor they require, elicit, and
extract across complex geopolitical, economic, and cultural interdepen-
dencies. According to Perlow, the unexpected visibility of the iPhone Girl
exposes "the complex forms of world-sharing by which the material, economic, and affective connections between producers and users of consumer
electronics get articulated." Her photo evokes the place of the body in the
material construction of the digital object, which subsequently serves as
the platform for the embodied performativity of the user. This performance
of use is particularly noticeable with mobile touchscreen devices (and will
be explored at length in chapters 3 and 4). "What happens," Perlow asks,
"between the regimes of corporeal discipline under which one set of bodies
assembles an iPhone, for instance, and those under which such devices get
used?"[103] Both regimes share a focus on productive labor bound in short,
repetitive strokes in seemingly small tasks involving complex forms and
connections. However, while the labor of assembly is seen as tedious and
menial, the labor of use is framed in fantasies of liberating performance,
supported by the rhetoric of marketers, journalists, and scholars alike.

The emphasis on gestural techniques characterizing interfaces of contemporary media nevertheless subtly reinscribes the historical, corporeal
discipline of equipment assembly into consumer use. It allows actions
based in the Taylorist scientific management of bodies and behavior for
efficient manufacturing to pass as gestures of freedom and play. "Consumer
electronics constrain embodiment in an array of contexts, though with
uneven severity," Perlow asserts. "By this view, the idealization of gestural
interfacing as an escape from such discipline—as a more organic, spon-
taneous, or natural way of interacting with machines—elides the bodily
suffering that repeated gesture itself occasions in the scene of production
and elsewhere."[104] The gesture of consumption seemingly effaces the earlier
one of production through what Perlow calls "aspirational affect," even as
the user's action may bear traces of the preceding movement, both in its
choreography and its consequences of component assembly (in this case,
the user's linking of data in the network).

So the mill girl is followed by the transistor girl, who is followed in
turn by the iPhone Girl. Jennifer Terry and Melodie Calvert point out the

continuities between contemporary circumstances and earlier intersections of gender and labor:

If women are particularly associated with "low-end" appliances such as … sewing machines, to what extent does this depend on the historically specific siting of these machines in the home, or to their status as accessories of women's wifely and maternal duties? What do we make of the fact that … a dramatically high percentage of the growing labor force in computer-based information management industries is female? Or that women in Malaysia and other parts of Southeast Asia are the primary assembly-line producers of the silicon microchips, the vital elements necessary for the very existence of the World Wide Web?[105]

The historical trajectory outlined in this chapter suggests an evolution of the history of textiles and hand production in computing that moves from the surface—or exterior—in the Jacquard apparatus's configuration of punched-card programming into the interior of modern computing's woven cores and integrated circuit boards by means of labor associated with the home, needlecraft, and domestic production. This history and memory then resurface—in literal and figurative ways—in contemporary tactile interfaces, their visual aesthetics, and the language and description of digital operations. The entwined history of textiles and computing traced here demonstrates that in both cases the labor of women remains invisible.

This invisibility is not only a matter of the political economy of technological production, however. Wendy Chun has demonstrated that digital technologies have continually and progressively relied on a paradoxical visibility and opacity in constructing the mystery of the device and its processes. This is particularly true in the function of software—a concept Chun asserts is historically gendered as female—and its use as metaphor. "Software seems to allow one to grasp the entire elephant [of new media] because it is the invisible whole that generates the sensuous parts," Chun states. It is "based on and yet exceeding our sense of touch—based on our ability to manipulate virtual objects we cannot entirely see." Software becomes an unknowable employed as a metaphor for the unknown.[106]

Embodied interface physically grapples with this unknown. It presents metaphors of action and materiality for otherwise unseen, unfelt, inscrutable processes, which bring to the surface in corporeal performativity not only the optical, electronic, textual, and digital processes of code and networked communication, but also the historically encoded performativity of gender and labor explored in this chapter. Coder Alan Sondheim claims that "every more or less traditional text is codework with invisible residue; every computer harbors the machinic, the ideology of capital in the construction of its components, the oppression of underdevelopment in its reliance on cheap labor."[107] The format and actions of mobile touchscreen interfaces, in their correspondences to the precision and skill of such labor and their deeper historical connections to informal, gendered labor training, become an interstice where this residue sticks. The histories laid out in this chapter, with their patterns of overlap, repetition, and recursion, suggest that contemporary shifts in material and social conditions of digital media reach back both temporally and spatially into the body of electronics and computing itself. Correspondences emerge at the intersection of craft culture, textile manufacturing, and methods of digital production and practice based on gestural and behavioral differences embedded in gender differentiation. These intersections occur throughout decades of digital media design at the level of hardware and software, leading us through a string of interactive models. Yet, the earlier textile forms of computing's hardware interiors come to the surface in a striking, very material way in the mobile touchscreen. The iPhone capacitive screen, which relies on electrical conductivity for converting finger contact and movements into commands, was designed with a fine mesh of wires just under its surface, echoing the grid of interlaced wires that formed the core memory of most mainframe computers. In describing first-generation iPhones, Jon Agar explains, "Sitting on the glass [under the cover lens] is a nearly invisible grid of fine electrical wires. ... The lines are about a millimetre apart. One line carries an electrical charge, while the other detects the slight disturbance caused in the electrical fields as your finger, acting like a weak capacitor, swipes the screen."[108] Here the touchscreen functions through variations of electrical impulses on the micro-fine wire grid, just as the core memory plane did. Now, however, those impulses are activated by the fingertip and the small, repeated strokes it makes across this base fabric, like a needle in embroidery, to build up small bits of data into a meaningful image.

## 2   Image Fabric

Digital images function much like fabrics, constantly woven and rewoven in their on-screen performativity. When an image is retrieved from a server or drive, its binary code threads through algorithms in microprocessors to be converted into the fine grid of pixels (short for "picture elements") visible through the mesh of diodes and cells that compose a screen's optical display. This image is therefore animated within the flux of the circuit—even if we perceive it as, say, a still photograph—as data are received, transmitted, and processed by the device and network. "On the computer monitor, any change to the image is also a change to the program; any change to the programming brings another image to the screen," Sadie Plant remarks. "This is the continuity of product and process at work in the textiles produced on the loom. The program, the image, the process, and the product: these are all the softwares of the loom."[1] Lev Manovich maintains that, in the conversion of image files to screen display, still images—as traditionally understood—no longer exist. Device and system operations merely suggest stability in the image. "It is only by habit that we still refer to what we see on the real-time screen as 'images.' … A static image … is no longer the norm, but the exception of a more general, new kind of representation for which we do not yet have a term," he stated in 2001. More recently, he has described such processes as "software performances," explaining that "what we are experiencing is constructed by software in real time."[2]

Images are increasingly the base material through which we interact with networks and each other. Perhaps because they tend to circulate in feeds, streams, and flows, we often consider them less as individual entities complete in themselves and more as components within larger series and constructions. This is as true for images in social media as those found in other network aggregators and search engine displays. In other words, images not only do not exist as stable physical objects, but also no longer function as isolated documents; rather, they serve as flexible components in patterns produced within the network of digital processes in and across devices and communication infrastructures.[3]

These properties of networked digital images are considered in this chapter through several vectors. By exploring the contemporary screen-image relationship of mobile, networked media in terms of the loom frame and fabric, it reconsiders shifting and sometimes antagonistic understandings of the photograph in regard to circuits and processes. It also explores important correspondences between the digital image's textile-like properties and photography's relationship with textile culture and computing in the mid-nineteenth century. In the earliest moments of their histories, photography and computing found a logical (if unexpected) point of contact in textiles, particularly lace, as an interface that rendered each medium visible to the other. Like photographs and mechanical calculations, textiles such as lace represented the concretization of abstract ideas, paradoxically appearing as an object of cultural production, yet resisting comprehension in its totality.

This legacy may inform approaches to networked digital images. Debates over the properties, status, and purpose of the digital photograph have frequently noted the handicraft conditions of its forms and uses, presenting it as a textile-like element of the manifold assemblages of networked culture. In some ways, digital imagery returns to the lace experiments of early photography by representing the complexities of human cultural production rather than reinforcing the mythology of a natural indexicality that loomed large behind the theorization of photography's ontological specificity for a century and a half. This shift is supported through digital image culture's recuperation of multiple aspects of textile production in the creation and circulation of networked visual representations. Accordingly, while theorizing the underlying processes behind the performance of digital screen visuality in terms of textile production, this chapter also considers contemporary processes of making, displaying, and circulating digital images. Such processes are based in textile piecework techniques including image stitching and image quilting, as well as in common interfaces and practices including infinite scroll software and image filters functioning across platforms such as Instagram, Snapchat, Twitter, Tumblr, and Flickr. As networked digital media are increasingly considered image-driven media, such affinities between the image and textile culture can only gain greater significance in shaping our understanding of everyday communication practices.

## Silk and Lace

The origins of photography's relationship to digital culture are not in Steve Sasson's 1975 invention of a camera for Eastman Kodak that used a charge-coupled device and magnetic memory to take and store a picture. Nor are they in the photo-based computing applications Vannevar Bush described to *Atlantic Monthly* readers in 1945 in "As We May Think."[4] Rather, they extend back to the beginnings of both technologies in the first half of the nineteenth century when they crossed paths in the friendship between Charles Babbage and fellow Englishman William Henry Fox Talbot. One of a number of inventors of photography that includes Louis-Jacques-Mandé Daguerre and Hippolyte Bayard in France, Talbot established the negative-positive process that remained fundamental to most types of photography until the digital turn of the late twentieth century. It was over breakfast at Babbage's home in 1831 that Talbot may have first considered the possibility of making images with photosensitive chemicals. On that morning the astronomer John Herschel demonstrated to Talbot and Babbage the application of platinum salts in making impressions from light.[5] Talbot, however, would later recall that his first thoughts concerning making images through photochemistry occurred during a more suitably picturesque moment while sketching in Italy in late 1833. As it happens, he was on his Italian tour during the same period that Babbage appears to have first given serious thought to his Analytical Engine. In any case, each man conducted rigorous experiments and made numerous adjustments to his respective invention in 1834 and by 1838 most of the fundamental elements of the engine and Talbot's "photogenic drawing" were in place.[6]

When Talbot made his invention public in 1839—following Daguerre's unexpected announcement of the direct-positive "daguerreotype" process—he turned to Babbage (and Herschel) to share his results, sending regular samples of his process and its improvements to Babbage from May 1839 into the 1840s. For Babbage's part, he demonstrated the Difference Engine to Talbot and informed him of progress on the Analytical Engine.[7] Babbage would show Talbot's latest improvements to the photographic process at his regular evening gatherings of intellectuals and scientists at his home in

London, where he also displayed and sometimes operated a working model of the Difference Engine. Occasionally participating in these social events, Talbot wrote to his wife in February 1840: "My pictures had great success at Mr. Babbage's last night,"[8] and it was possibly at one such gathering in 1844 that Babbage lent some of Talbot's photographs to Ada Lovelace's mother.[9]

In demonstrating and promoting his image-making invention early on, Talbot was reluctant to rely on the faint scenes produced with the camera obscura. Their accuracy and utility seemed limited, since they reversed light values and Talbot had not fully accepted the viability or desirability of printing "positives" from these "negatives." Instead, he circulated contact prints made from flat objects placed on writing paper that had been soaked in his photosensitive chemicals. These included etchings, engravings, plant samples, and pieces of fabric.[10] Lace, like that produced on Jacquard-equipped looms, was among Talbot's first and most important photographic subjects (figure 2.1). He made successful prints of lace as early as 1834. In 1839 these prints would demonstrate to an eager British public photography's capacity for precise and complex reproduction.[11] He sent a contact print of two lace samples to Babbage in May.[12] Two months later, in his first public exhibition of images produced by his invention (organized for the meetings of the British Association for the Advancement of Science in Birmingham), Talbot included over twenty contact prints from lace, muslin, and calico. The first audiences for his prints were drawn to those depicting lace, rather than other subjects, as the most striking and attractive examples of the medium.[13] While he included images of plants and fruit in an album he sent to Queen Victoria, one of her ladies-in-waiting reported back that "the Queen was more struck by the exactness of the ribbon than the beauty of the ferns and grapes ... the gauze ribbon she said was very curious."[14] Talbot would claim that such textile prints could be mistaken for fabric. One correspondent confided to him, "I was so completely taken in by your lace-picture, that ... I was actually handing it over to [my wife] as a lace-pattern, intended for her."[15]

It appears that for both maker and viewer, these prints from textiles were more than simply a convenient means of demonstrating the capacities of photography. While it is understandable that contact prints may have had a greater impact than the softer, less recognizable images typically produced by the camera at that time, lace's success above all other subjects pointed

**Figure 2.1**
Contact print of lace, by William Henry Fox Talbot, made before December 1845.
Digital image courtesy of The Getty Open Content Program, 84.XM.478.14.

to a contemporaneous interest in the complexities of manufacture and calculation as well as photography's potential application in these. Within days of revealing his invention, Talbot arranged for photographic prints of lace to be shown to textile manufacturers in Glasgow. The following year, he was asked to exhibit prints of lace and other fabrics at the Frankfurt trade fair.[16] Britain was a major textile exporter in the 1830s and 1840s and Glasgow was a center of its production, harboring over a dozen lace factories and accounting for as many as eight thousand handloom weavers and an equivalent number of factory-based textile workers. As W. Hamish Fraser states, although the power loom had been introduced in the 1820s and brought more women into the textile industry, the number of handweavers continued to grow in Glasgow, "concentrating on fine weaving and on the complexities of pattern with which the powered technology could not cope."[17] Talbot's image reproduction process seemed to indicate a potential means toward further mechanization of the industry, where one day both skilled weavers and punched-card technologies might be superseded by photography's ability to create a template from which products could be manufactured relatively quickly and accurately. As Douglas Nickel remarks, "Behind Talbot's presentation of lace images lay the development of the machine-made lace industry in England."[18]

Geoffrey Batchen believes the prints of lace that Talbot showed Babbage also served as a visual means of aligning photography with computing. In the prints' stark contrast between light and dark, Batchen claims that "Talbot rendered the world in binary terms, as a patterned order of the absence and presence of light."[19] Talbot included a print of lace in his multivolume treatise on photography's potential uses, *The Pencil of Nature*, specifically to demonstrate the principle of negative and positive, which would become the foundation of both photography and digital computing. As Batchen states, "This is a photograph not so much of lace as of its patterning, of its regular repetitions of smaller units in order to make up a whole ... Here was mathematics made visible."[20] Other early photography researchers, such as Bayard and Herschel, also turned to lace and fabric as their interface with photographic processes. Like later digital imagery, these contact prints were made without a camera, as though they were the starting point of image construction.

Talbot may have envisioned his invention as a means to one day replacing Jacquard's punched-card apparatus and similar industrial systems of representation and production.[21] At the very least, if, as Ada Lovelace claimed, "the Analytical Engine *weaves algebraic patterns* just as the Jacquard loom weaves flowers and leaves," then Talbot's invention became the mechanism for rendering visual correspondences between the complex patterns of natural objects such as leaves and flowers, on the one hand, and lace, textiles, and other manufactured fabrics, on the other. All seem to be reproduced in the photograph with a mathematical exactitude—one can count the cells or loops against the original. Indeed, one of Talbot's earliest camera-based images, "Latticed Window (with the Camera Obscura)" from 1835 was notable for its fine reproduction of the patterned divisions of its subject. "When first made," Talbot's handwritten caption reads, "the squares of glass about 200 in number could be counted, with help of a lens." In other words, one of the founding images of photography was an exercise in pattern, numbers, and the subvisual division of image surface that later became standard with cathode-ray and light-emitting diode (LED) screen displays.

The same principle would guide Talbot's invention of half-tone printing in the 1850s. In seeking a viable method for reproducing photographs in ink from metal plates—a significant step toward high-volume image production—Talbot found that images made from photosensitive plates suffered a severely restricted tonal range when the plates had been created directly from a translucent photographic positive. He returned to the textiles instrumental in his invention of photography to discover that by first making contact prints of gauze—or other finely woven fabrics—on a plate (figure 2.2), then exposing it to the desired photographic image, he could create a subtly textured surface that rendered a wide tonal range in ink

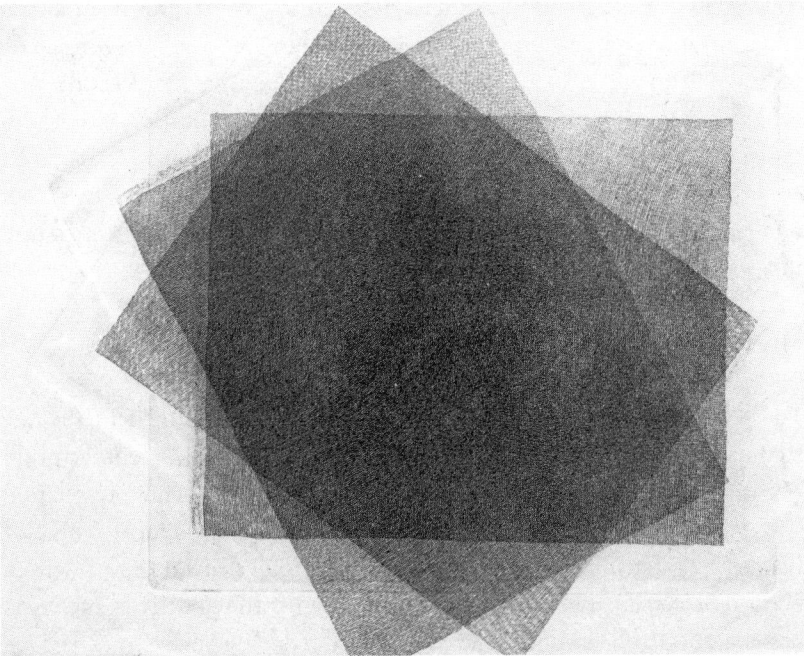

**Figure 2.2**
Photographic engraving of three sheets of gauze crossed obliquely, by William Henry Fox Talbot, made about 1852–1857. Digital image courtesy of The Getty Open Content Program, 2004.88.1.

printing. This principle of a standard, evenly spaced pattern of cells upon which any variety of images can be produced is behind nearly all modern image production and display systems. The fine honeycomb of the LED screen descends directly from these textile-based image systems. "These things, which I believe have not been heretofore used in the fine arts, I would denominate photographic screens or veils," Talbot wrote in his 1852 patent application for the process, adding that a single screen could be used for multiple plates.[22]

For digital technologies, the most unexpected and far-reaching implications of Talbot's photographic inventions are in the photolithographic printing of microchips, which resorts to the simple methods employed in Talbot's experiments with lace. Essential to the architecture and design of the microchip, photolithography uses minute, intricately cut metal masks—the contemporary industrial equivalent of a piece of lacework—to make direct contact or projection prints of circuit patterns on a chip's silicon surface.[23] Here Talbot's analog photographic process finds its industrial application over a century after his first inquiries in the Glasgow textile industry, through a virtually invisible material object that becomes the substrate for another form of image making. This application is not photography as the medium has been popularly understood, of course, but it comes exceedingly close to photography as it was first theorized by Talbot, as a process not only for making images, but also for making things from images.[24]

## Screen Looms

The "new kind of representation" that Manovich identified in digitally stored, screen-accessed visual documents at the turn of this century prioritizes the electronic device and its interface as critical factors in our relationship to, and understanding of, nearly all contemporary images. This shift has led to a fetishization of the screen as an object in itself, a phenomenon reflected in the conspicuous marketing of the aesthetics of screen-based mobile devices. The dual trend toward reducing device thickness and increasing screen size (as percentage of surface area) has fueled this material aesthetics. Phones and tablets may strike us now as nothing but screens: unadorned, millimeters thick, flush-edged, devoid of buttons, lights or any other distractions from their surface and shape.[25]

By privileging the screen in this way, the object is reinforced while the image it displays must adapt. Filling the screen surface with the image has become paramount, even when this risks alterations to the image's original aspect ratio or its internal, formal relationships. The image becomes eminently convertible in contemporary interfaces, there to be stretched, compressed and rotated in conforming to all manner of screen frames in a proliferation of formats and dimensions developed for any number of devices and browsers. The image on touchscreen mobile devices can be adeptly pinched and pulled by fingertips. Default settings in operating systems, programs, or applications may automatically adjust the image on these displays to stretch visuals across the entire screen like a piece of knitted fabric. In turn, the activated image reinforces the screen and its frame by responding to touch commands and displaying in ways that heighten a user's awareness of the frame (by rotating ninety degrees when the frame is turned, perhaps, or zooming into—or out of—full-screen mode).

Anne Friedberg explains that "like the window, the screen is at once a surface and a frame—a reflective plane onto which an image is cast and a frame that limits its view."[26] In exploring this definition, Friedberg cites Jacques Derrida's theorization of the "parergon" as a critical step. It is worth exploring the concept of the parergon here, as it bears consequences for the changing function of the screen frame and its edge in relation to textile culture. Working from Immanuel Kant's use of the term in *Critique of Judgment*, Derrida describes the parergon as "that which is not internal or intrinsic, as an integral part, to the total representation of the object but which belongs to it only in an extrinsic way as a surplus, an addition, an adjunct, a supplement." The frame serves that function and remains as important to the functioning and understanding of the image-as-image as it does to the visual information the image contains. "The parergon inscribes something which comes as an extra, *exterior* to the proper field," Derrida states, "but whose transcendent exteriority comes to play, abut onto, brush against, rub, press against the limit itself and intervene in the inside only to the extent that the inside is lacking. It is lacking *in* something and it is lacking *from itself*." He elaborates: "Parerga have a thickness, a surface which separates them not only (as Kant would have it) from the integral inside, from the body proper of the ergon, but also from the outside, from the wall on which the painting is hung, from the space in which statue or column is erected, then, step by step, from the whole field of historical, economic, political inscription in which the drive to signature is produced."[27]

As Derrida points out, the parergon informs the ergon (the work) through its own formal beauty. If it is not formally beautiful it may descend into distracting adornment.[28] In certain circumstances of the frame-image relationship enacted by today's screens, the relationship has been reversed. Not only is the screen device's prevailing austere form far from adornment, recalling the simple frames of the twentieth-century home weaving kits described in chapter 1, but one could point to the image it houses as fulfilling such an aesthetic function. The frame can be more formally beautiful when juxtaposed against the potentially "deformed" image that it contains. It is this deformation, in fact, that can work to emphasize the formal beauty of the frame. If the contemporary digital image is a simulacrum that need not refer to any existing or perceived "original," as appears to be the case when it circulates and recirculates through the vast sprawl of today's telecommunications networks, then it is available to (and should) be modified to enhance the aesthetic value of the material object before us.

Significantly in the current context, Derrida's analysis stems from Kant's exploration of the garment as a parergon on religious or mythological figures, as with sculpted drapery on statues. Derrida archly asks, "Where does a parergon begin and end. Would any garment be a parergon[?] ... What to do with absolutely transparent veils[?]"[29] From this musing he jumps to an image: Lucas Cranach's *Lucretia* of 1533 and her transparent veil. This invocation of the textile unexpectedly intersects Derrida's deconstructionist pursuit with the analytical philosophy of Stanley Cavell. Cavell similarly grapples with the screen's frame, working with the movie screen as his principal object of inquiry. Cavell theorizes that the screen in cinema produces a frame more closely related to the frame of a loom or a house than to a picture frame. The screen acts as a "mold" or "form" for the image, rather than a border. "Because it is the field of a photograph, the screen has no frame; that is to say, no border," Cavell claims. "Its limits are not so much the edges of a given shape as they are the limitations, or capacity, of a container. The screen *is* a frame; the frame is the whole field of the screen—as a frame of film is the whole field of a photograph." When projected onto this surface, Cavell explains, "successive film frames are fit flush into the fixed screen frame [which] results in a phenomenological frame that is indefinitely extendible and contractible, limited in the smallness of the object it can grasp only by the state of its technology, and in largeness only by the span of the world."[30]

The cinema screen has no frame, as Cavell argues, because it *is* frame. Frame and image bear a one-to-one correspondence; where one ends, so must the other. In the cinema, for example, it makes no sense to speak of screen space that is *not* image space. With mobile digital media, however, this is not only possible but it also points to a relatively common occurrence. Today's screen frames are closer in form to the picture frame than they are to the cinema screen, yet they may act as molds or loom frames, fitting (refitting, custom-fitting) the image to their form. They therefore function outside these opposed actions. If the picture frame directs attention to the interior, and the cinema (and traditional television) screen acts as a window on an expansive exterior world, the new frame-image relationship directs attention to the frame as delimiting edge and shaping container.

Despite drastically different theoretical perspectives, reading Derrida and Cavell together introduces an unanticipated corollary pairing of frame and fabric. The veil or drapery as a parergon, the frame as a loom—each represents a way of envisioning the relationship between the edge and the surface of the image that survives in our interactions with contemporary digital devices. There is a continuing, constant tension produced between hardware and software, casing and screen, and frame and image, in our embodied experience of the network. Under these circumstances, the image as a fabric to be manipulated becomes a significant and helpful metaphor that plays out in the visuals, movements, and products of interface.

In a digital device's processing, the base material of data and code—including all that contributes to the look and functioning of the interface—is continually interlaced. The central processing unit converts data input to information output by reading and computing, essentially converting sequences of ones and zeroes into a meaningful and useful object for the user. If data and code function primarily as a single stream or thread, of ones and zeroes, then digital devices function in ways very similar to knit-fabric production. Whereas woven fabrics rely on the intersection of two threads—the warp and the weft—to build their form, knitted fabrics can be rendered from a single thread (figure 2.3). It is true that the binary aspect of code operates in ways similar to pattern weaving, as pointed out in the discussion of Jacquard's apparatus at the start of chapter 1. A byte of eight

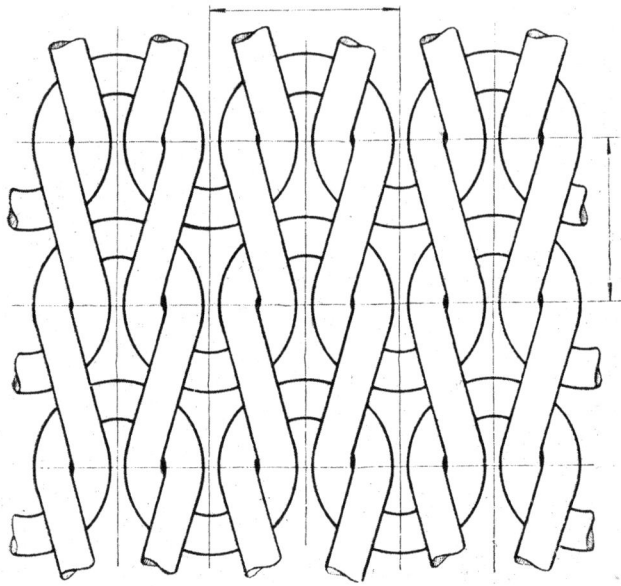

**Figure 2.3**
Diagram of a single-knit structure. Drawing by Elkagye.

bits, for example, is comparable to a pattern for a weave on an eight-thread warp, where the weft can travel either over or under each of the warp's threads with each passage of the shuttle, producing any of 256 combinations. Eight-bit color graphics works similarly, allowing 256 color choices for each pixel. However, if a digital graphics file is processed as a single stream, data and code may function more like a long, intricately dyed thread that, when *knitted* in the proper order on the loom frame of the screen, produces an intelligible image.[31] Every time this data and code are accessed, whether minutes or years apart, it is knitted into the same image. However, just as handlooms come in different shapes and sizes, screens come with different specifications and aspect ratios. As the same networked image will be knitted on different screen frames, there can be distortion, aberrations, or discrepancies across instances and frames. Common image-display settings for operating systems, programs, and applications, such as full-screen, wide-screen, stretch, and zoom modes therefore depend on a contemporary perception of flexibility in the image as a material that, like any knitted fabric, can be stretched along its width or length to accommodate the shape of the underlying object.

This understanding of the digital image as inherently different from previous types of images (e.g., printed photographs or celluloid negatives) explains the image's frequent deference to the varying conditions of the screen frame and supporting software. It also sheds light on the initial fears surrounding digital photography in the 1990s as a post-photographic medium that jeopardizes the presumed evidentiary value photography accrued through epistemologies formed around earlier photomechanical methods of image-making. The digital image—as image—is always unstable in its absolute dependency on the performance of the electronic device that builds it from binary code into a meaningful visible display. In its digital origins it is nothing more than the on/off code also found in the up/down warp thread positions of the weaving process. In its visibility on-screen it is always enmeshed in the relationship between software, processor, and screen properties. And, more and more, it is further entwined in the wider networks and patterns of information that radiate beyond the device.

### Seamlessness

Through photolithographic microchips, analog photography is physically inscribed into all aspects of today's mobile media devices and networks. Most digital images therefore maintain a seldom-acknowledged material base in analog photography. Perhaps it is no coincidence that Fairchild Semiconductor, which pioneered semiconductor technology, was a subsidiary of Fairchild Camera and Instrument.[32] The integrated circuit is a direct descendant of photography optically, mechanically, chemically, and economically.

This is not how the story typically goes, however, in discussions of photography and the digital turn. If we can consider the contemporary relationship between digital images and screen devices as analogous to that of fabric and loom frame, and find worthwhile correspondences in the discourses of photography and computing in the mid-nineteenth century, the pivotal episode between these two moments is the cultural and technological shift from analog to digital photography at the millennium, and the alarm it provoked concerning the ontology of the image and associated epistemologies of truth and objectivity. Placing the "computational" image composed of binary code in opposition to the photographic and electronic image, Edmond Couchot warned as early as 1988 that "a new image is emerging, belonging to a fundamentally different symbolic economy."[33] Batchen offered a more tempered analysis. "Given the advent of new imaging processes, photography may indeed be on the verge of losing its privileged place within modern culture," he wrote in 1997. "However, any such shift in significance will have as much to do with general epistemological changes as with the advent of digital imaging."[34]

William J. Mitchell, an early critic of the consequences of the digital turn for the status of the photographic image, saw the raster grid of the screen and the pixel-based image as the fundamental tropes of digital photography, replacing the camera and film of analog processes. He situated digital imagery's rupture from earlier ideas of the photographic in a schism between the continuous and the discrete. Analog technology was continuous for Mitchell, while digital was discrete. The raster's diode or the image's pixel represents the basic unit of the image. The chemically based photograph has no such basic unit. "Images are encoded digitally by uniformly subdividing the picture plane into a finite Cartesian grid of cells (known as *pixels*) and specifying the intensity of color of each cell by means of an integer number drawn from some limited range," Mitchell explains.[35] As has been argued earlier in this chapter, a digital image can be conceived as continuous, but its continuity rests in the performance of its construction at the time of display, rather than initial production. It is an idea of continuity quite different from Mitchell's, to be sure, but it could be said to underlie current perceptions of the performativity and utility of the digital photograph.

This aspect of the digital photograph as an image built from strings of numbers may seem straightforward, but in the popular conversion to digital photography at the start of the twenty-first century it was seen as an ontological shift for the image and visuality. The image was no longer a "direct" trace of optical phenomena, documenting what occurred independently in the physical world through the indexicality of photochemical processes. It was instead the occurrence's conversion into another language or register that—at its base—bore no formal resemblance to the physical events to be depicted. "The pixel comes first from language, a formal language certainly, but nevertheless a language," Couchot explains. "It does not interpret any preexisting reality."[36] Indeed, one could say that the digital photographic act leaves no recognizable trace of an event in physical reality, thereby threatening any remaining claims for the camera and photography as "transparent" media. Batchen called it "a pliable sequence of digital data and electronic impulses … all about the reproduction and consumption, flow and exchange, maintenance and disruption, of data."[37] Indeed, the systems and operations basic to digital image production initially were considered so radical a departure from traditional photographic technologies that, for some, even using the terms "photograph" or "photography" was considered misleading. "Photographic image" and "digital image" were offered as more appropriate alternatives.[38] Others simply declared the "death" of photography, making "digital photography" paradoxically "post-photographic."[39]

The digital image was construed as a threat because it upset the false, but enduring, sense of stability in the photographic as natural and truthful. "A photograph is fossilized light," Mitchell claims. "It is a direct physical imprint, like a fingerprint left at the scene of a crime or lipstick traces on your collar. … There is no human intervention in the process of creating the bond between photograph and reality, this apparent Kryptonite connection to the referent: it is automatic, physically determined and therefore presumably objective."[40] In essence, the analog photograph had represented an important instrument of power. John Tagg has argued against any "phenomenological guarantee" in analog photography, however. "We have to see that *every* photograph is the result of specific and, in every sense, significant distortions which render its relation to any prior reality deeply problematic." He explains, "That a photograph can come to stand as *evidence*, for example, rests not on a natural or existential fact, but on a social, semiotic process" of institutional practices and historical relations.[41]

Fred Ritchin vividly describes the early fears digital photography inspired in some, recalling the moment he had gazed at advertisements while riding the New York subway and realized that the images could be entirely digital productions: "The question for me was … whether in fact the person or scene had even existed. I began to sweat, unsure as to whether this entire system of referents was functioning. Seeing certainly was not believing, and the photographs seemed to represent openings into an alternate universe synthesized according to discordant goals."[42] Looking back at Ritchin's and other circumspect accounts of digital photography of a decade or two ago, what emerges as the most feared aspect of the new digitally based visuality is its "seamlessness."[43] As a fabrication easily produced from joining multiple image fragments in the "synthesized" universe described by

Ritchin—the digital image should evince the seams of its making. That it can be assembled like textiles but may not betray the joints of that assembly becomes its root danger. Conventional analog retouching typically left traces—or seams—detectable by the astute viewer. Digital photography editing would not, or so it was argued. The conventional photo became untrustworthy the moment it took on these traits of a new sort of fabric, a piece of handicraft where the needlework is invisible despite being an "electrobricollage."[44]

With an emphasis on loss of indexicality as the key factor, many late twentieth-century accounts of photography's move from analog to digital processes missed or ignored the significance of the shift from making and circulating paper-based images to relying on screens for both production and consumption. As printed photographs have become less and less common, the photograph has transformed into a performance not just at the moment of production, but at its viewing. As data processed for screen display, digital images are constantly assembled and reassembled by processors as they are stored and accessed. In accessing networked images, when bandwidth is limited and processing speeds lag this assemblage will take place slowly on-screen, much to the viewer's frustration. It can take form through an integral, progressive loading, which displays a low-quality version of the entire image that gradually gains greater resolution. However, files may also load line by line at full definition, beginning at the top, performing a visualization that resembles the knitting or weaving process. This was a common occurrence when accessing online images in the early years of the Internet, almost becoming the aesthetic of the online image and a visual sign of the Internet as a continuous flow of information into the computer. Even today, while JPEG-formatted images may tend to load integrally, those in PNG, GIF, and other formats sometimes load linearly, potentially producing varying experiences of the image as it is constructed on-screen.[45] A single image stored in multiple file formats, therefore, can produce different screen performances, depending on which format is accessed, even if the files are of comparable image quality.

### Stitching and Quilting

While knitting and weaving serve as apt metaphors for understanding the functioning of digital devices, particularly as they concern the production of individual images, sewing and quilting are more appropriate when considering the networked imagery typical to contemporary mobile media use. Networked images—whether shared on a social media platform or stored for personal use through cloud computing—differ from those stored directly on a device in terms of access, processing, and display. While access and processing may remain hidden from the user (except when things go wrong), the aesthetics of networked image display, such as infinite scroll, have become central components to our understanding of visual culture today. If, as Ritchin claims, photography now consists of "creating discrete and malleable records of the visible that can and will be linked, transmitted, recontextualized, and fabricated," then we must consider the Internet and networked computing as integral parts of our understanding of the photograph.[46]

A common form of the networked image is the "stitched" image. Image stitching is a key component in creating the networked visual environment, contributing to spatial continuity within panoramic images, 3-D image spaces, game worlds, GPS interfaces, and locative apps. Image stitching can create a single image of a scene from a series of images bearing fragments of the scene taken from one vantage point. This requires the photographer to take a sequence of shots, being careful not only to capture every zone of the scene, but also to ensure that elements on each edge of any given shot will repeat in the shot of the neighboring zone. These overlaps of elements provide the extra material for "stitching" the composite. When the work of joining images is done using general photo editing software rather than through programs made specifically for the task, the technique is known as "hand-stitching."[47]

The two basic operations in image stitching are alignment and blending. After identifying common elements as registration points to align two adjacent images, a stitching program or app will blend the opposing edges of the two images into each other. One patent application describes the process as "determining a set of edge position coordinates[,] … generating a set of possible matched solutions … [and] blending intensity values of pixels" from one image with the "corresponding" pixels of the other, thereby "smoothing the intensity discontinuities between images."[48] By combining material found at the edges of each image to produce an undetectable bond between separate parts, the algorithms of image stitching resemble traditional techniques of darning and patching. Blending pixels need not necessarily change the color values of extant pixels, but rather it arranges them in a pixel-by-pixel patchwork that blends the edges where the images conjoin.[49] Similarly, a darn uses thread to intersect with extant fabric in ways that blend the worn pieces. "Do not trim off the frayed or worn edges" of fabric, warns one early twentieth-century guide to mending clothes. "The unevenness around the edge, which these frayed ends create in the process of darning, helps to make the darned place less conspicuous."[50]

The processes behind image stitching were developed in the 1990s and described at the time as "merging," "layering," "assembling," "combining," and "superimposing" to create a composite.[51] With the Photomerge options introduced in Adobe Photoshop Elements 1, however, "stitching" became a common way to describe this process, even if the software systematically employed variations of the verb "combine."[52] "Nature photographers have created panoramic photographs for many years by taping pieces of film together. ... The [Elements] program gives you a way to quickly stitch several images together into one solid sweep on an image," explains a 2001 manual.[53] The software's functioning was even illustrated at the time through the metaphor of needle and thread.[54] The arrival of this method for making composite images was sometimes enthusiastically received (a Photoshop Elements handbook considered it "the one feature that just might make long-time Photoshop users want to plunk down a hundred bucks to buy Elements"[55]), even as it placed the terminology of home handicraft into the discursive register of the panorama, or those views that historically seemed to transcend the frame of media to produce a sense of all-encompassing and unlimited "natural" vision. The view was paradoxically "pieced together (stitched) into a seamless panorama."[56]

Building a view from overlapping photographs arranged both horizontally and vertically is sometimes called stitching or making a "mosaic," but it is also known as "quilting." While automatic stitching of linearly aligned photos was available by 2001 in Adobe PhotoDeluxe, Adobe Photoshop Elements, MGI PhotoVista, and similar photo editing suites, quilting often had to be done by hand. Automatic quilting features emerged a few years later, allowing not only for the stitching of multiple photographs, but the stitching of patterned fields from a single, repeated photograph. This has been particularly useful for creating larger, patterned backgrounds, such as wallpaper, from a single image or fragment. In textile assembly such as suit- or dressmaking, any use of patterned fabric requires close attention to the relationship of conjoining parts both in orientation and pattern continuity. While this is also a concern in image stitching, as when architectural elements are conjoined in a composite, it is central to image quilting. In image quilting, a larger field is built up from a single image or by "stitching together small patches of existing images."[57] Like fabric, what the image depicts commonly contains a strong pattern or texture, such as brickwork or a plot of grass. In the case of brickwork, the image would appear highly structured and regular. With grass, however, the structure would appear random, with every blade differing from the others, even if the overall effect is even and balanced. Stitching and quilting are instrumental to digital image interactivity. They contribute to the sense of continuity and endlessness the viewer feels in exploring image-based digital environments, from open world games to apps such as Google Street View.[58] These images may be constructed from photographs, computer-generated imagery (CGI), or any combination of the two.

## Switching and Scrolling

The stitched and quilted aspect of digital imagery extends into the processes of networks and interfaces. It is replicated, for example, at the moment networked images are accessed through a device. Despite—or perhaps because of—their size and reach, contemporary digital networks are pieced entities. Information, such as images, that seemingly comes to us whole under normal device and network operations in fact arrives in pieces, having been broken into segments that travel independently through the network to be reassembled by software on the receiving end. Segments arrive asynchronously as "packets" to be rearranged and opened in their proper order before being presented through the screen (and speakers, in the case of audio content) as a cohesive unity. This system of networked delivery is known as "packet switching" and remains a key component of the fast, efficient transfer of data through distributed networks. "Switch fabrics" of silicon chips that house gridded circuits are placed at points throughout the network to facilitate these transactions. Based on telephony, a switch fabric allows bits of data to travel along the fastest path, adjusting to traffic fluctuations. At the final destination, the received pieces are sutured back together. In other words, switch fabrics act as stable textiles within the mesh of the network.[59]

If we pull back from the individual image to examine the effects that networked images produce in their multiplicity, new forms emerge. With digital image circulation, the archive and the album have been replaced by the stream and feed. Photography-centered platforms have moved away from album-based structures to photo streaming structures, where images are followed by more images that become the latest row in a continuously stitched and restitched textile. In describing the digital photograph as an "algorithmic image," Daniel Rubinstein and Katrina Sluis claim that "online there is no point at which the image ends; rather, there is an endless succession of temporary constellations of images, held together by a certain correlation of metadata, distribution of pixels or Boolean query."[60] Social networking sites and apps, with their heavy reliance on image making and image sharing as a means to interaction and participation, form a key area of this activity. On Instagram, Snapchat, Pinterest, Tumblr, and similar platforms, images are rarely presented individually but rather appear in tight, dynamic rows, columns, and grids.

The programming and display technique underlying this shift is known as infinite scroll. Similar to packet switching, infinite scroll relies on asynchronous web applications (often employing JavaScript) that allow data to be sent and received between the device and the network without disrupting the display or functioning of the interface. Sometimes called "endless scroll," infinite scroll allows dynamic loading of additional content as a user scrolls a webpage or app. Through data caching and event buffering, it reduces or eliminates page breaks within the interface. When a user scrolls to the lower limit of content that has been loaded for display, additional retrieved content is automatically appended and further content is requested from the network. The software behind infinite scroll was developed in early 2006, little more than a year before the iPhone drove the global shift to touchscreen mobile devices.[61] Infinite scroll's emphasis on scrolling instead of clicking makes it particularly suitable for tactile interfaces, where finger strokes allow faster and smoother movement around a page than is customary with a touchpad or mouse. It also suits the vertical screen orientation more common to phone use, allowing for long streams of content to be perused. Infinite scroll has become an effective means of displaying frequently updated and augmented chains of information, as might be found on social media platforms. As one blogger notes, "Users are aware that they won't get to see everything on these websites, because the content is updated too frequently. With infinite scrolling, social websites are doing their best to expose as much information as possible to the user."[62]

There has been debate about the effectiveness of such interfaces, however.[63] Because more and more data is loaded on a single page as a user scrolls down the content, infinite scroll can become unwieldy and frustrate a user's attempts to interact with displayed data. Dimitry Fadeyev claims that, for example, infinite scroll is effective not when "the user is *searching* for something in particular within the list of results," but rather when "the user is scanning and *consuming* the flow of information."[64] Although Twitter's feed of short text messages is often invoked as an effective and sensible implementation of infinite scroll, the interface also has enjoyed considerable success with image-based sites and displays. Text-based data often requires closer attention to differentiate content, while images may be browsed quickly and in groups. Given this difference, Google has adopted an infinite scroll format for displaying image search results, while adhering to a conventional paginated format for general searches.

On touchscreen phones, infinite scroll is sometimes a single-column display tailored to the screen's vertical orientation. The functioning and presentation of this single-column interface resembles chain piecing, a common, assembly-line approach to stitching blocks together while constructing quilts.[65] Also called "string sewing," chain piecing is a mechanical method for stringing the squares of a quilt "together on threads, like beads," before they are organized into larger structures of rows.[66] In chain piecing, squares are stitched together on the sewing machine in chains as long as the vertical length of the quilt. These pieces may be sequenced to produce a pattern, but they may also be sewn in a random order called improvisational or "improv" piecing. Improv piecing produces a quilt in which patterns are not predetermined but may emerge afterward. Relationships and patterns in this form of quilt making change over the course of the work's production. The relationship between any three pieced squares, for example, is prone to change when a fourth and fifth piece are added. The same performative, aesthetic parameters shape the experience of image-based, social media platforms such as Instagram and Snapchat. Part of the enjoyment of these platforms lies in identifying motifs, patterns, and themes as they emerge from the chain piecing of images. The stream of

**Figure 2.4**
Partial results from image search for "daisies," as displayed by Google search engine's infinite scroll interface, January 2017.

images inevitably produces a "conversation" across images that may or may not be evident to the maker or makers.[67]

Under other circumstances, especially on larger screens and horizontally oriented smartphones, aggregated image results are presented in tight formations of images that may be resized to produce uniform rows (figure 2.4). Here infinite scroll can be said to mimic the weaving process of a bolt of cloth being made by the even, horizontal passages of the weft thread with no definitive or required end. Just as the end of the bolt is ultimately determined by the length of the warp threads used, the end of an infinite scroll display can be linked to processing power or data limits. However, infinite scroll does not assemble a single piece, as is the case with weaving or the line-by-line display of a slow-loading image. Instead, it assembles multiple, visibly distinct components modularly into a coherent, patterned whole. "In these database-driven image systems linear narrative becomes subservient to the logic of computer-based data modelling," explain Rubinstein and Sluis. "The cinematic world of montage collapses into that of correlation—of similar size, similar patterning, similar tags, similar metadata, similar location."[68]

In the gridded composite built from scattered and independently created visual components, infinite scroll functions less like a woven textile and more like patchwork quilts of pieced blocks of fabric. Patchwork is the best-known type of quilt. Like infinite scroll it "involves assembling hundreds of individual patches side by side."[69] The pieced blocks of such a quilt may follow a standard, simple pattern, such as a grid of squares, but the squares may bear any number of visual and semantic relationships to each other. They may be identical, they may be alternating patterns of the same or similar design or color, or they may have no discernible relationship to each other beyond being squares of fabric placed in a sequence. In other words, patterns and relationships may be built into the grid or they may emerge only after the work is completed. Of course, oftentimes the organization of the squares may have some relationship for the maker of the quilt that may escape the viewer, and vice versa. Gridded displays of digital images common to website and application interfaces may work similarly. This is particularly pertinent to apps such as Google Images. Once an image is loaded into the app's search window, Google furnishes a gridded grouping of images under the heading "visually similar images." As the title suggests, the idea is that these images are related in their visual composition—shapes, colors, etc.—rather than in the textual clues attached to them, such as the websites on which they are located, their captions, or their file names.[70] In other words, recognition software offers grids of visually related squares. This model of organizing and understanding images, in addition to practices such as tagging (to be considered in chapter 3) open onto what Rubinstein and Sluis call the "undecidable" image. "The networked image is undecidable because the meaning of the image is not fixed to any specific event but to the progressive accumulation of a 'data shadow' that determines its visibility and currency in a range of situations."[71] This shadow of data represents the sum of the threads that can tie any specific image into multiple groupings of images in different display modes.

The textile culture aspect of these networked images emerges most strongly in the ubiquity of interfaces that fill the screen with images arranged nearly edge to edge while allowing—and even encouraging—additive processes that recall appliqué, overstitching, and similar needlecraft practices. Instagram has been a leader in this regard, which may explain its dominance as a photo-sharing site in the touchscreen era. Instagram distinguished itself when it was released in 2010 by adopting infinite scroll as its display mode when other photo-sharing sites (e.g., Flickr and Photobucket) still relied on page displays. It offered users an aesthetic approaching a needlecraft model in at least two ways: through a standard of square images that displayed as a perfectly fitting grid, and via filter options that could alter the hues and color temperature of each image square. Instagram's signature square aspect ratio diverged from the popular norm of "landscape" formats of photographic programs and mobile-device screens. As one blogger wrote when Instagram adjusted its app to require square formats, "the fact of the matter is that non-square images look crappy in the stream and they look bad in the web view, so it's trying to make a product that looks the best it can."[72] Given the prominence of the square image as a defining design element in Instagram's interface, the company's decision in 2015 to allow non-square formats drew strong reactions.[73] Once in the app, selecting an account will produce a display of thumbnail images neatly arranged with a thin white border around each of them. The photos are displayed in the order the account owner uploaded them. Selecting a thumbnail will display the corresponding photo individually with any attendant text or emoji. Selecting another account name or tag appended to the image produces a fresh display of thumbnail images either from that additional account or bearing that tag. In the case of the tag, the display brings together images from multiple accounts around a common event or sentiment, producing a themed scroll display that bears social and affective parallels with themed quilt making.

Instagram's principal competitor in image-based social media, Snapchat, offers a different interface experience. Meant to serve as an ephemeral statement in a real-time visual dialogue, most often Snapchat photo "snaps" are made and sent to a recipient who views them individually for a few seconds before they disappear from the device's memory. As such, the quilt making affinities of infinite scroll's multi-image displays did not apply to Snapchat until the 2016 addition of a "Memories" feature that allows archiving and organizing of images around a common theme.[74] Snapchat has embraced a craft approach at the level of the individual image, however, through drawing, filter, and effect options. These options rely on assemblage and modification strategies that relate to techniques of embroidery, overstitching, and appliqué. For example, drawing lines or letters on the image (figure 2.5) functions like freehand embroidery, which allows sewn threads to produce lines and images on a textile without taking into consideration the structure of the underlying weave (warp and weft in the case of textiles, pixel grids in the case of digital images). Just as appliqué traditionally involves "various elements of the design, such as hearts, flowers, leaves, birds, vases,

**Figure 2.5**
Snapchat photo employing the app's drawing feature, 2015. Image by Lane Fournerat.

[that] are cut from various fabrics and sewn onto the background fabric," Snapchat provides a range of emoji and symbols—including hearts—to be placed over the underlying image.[75] Though the images may be admired for only ten seconds or less, the number of editing tools available in Snapchat means an image may be richly embellished before being sent, suggesting that the act of constructing these images by hand may be as enjoyable as sharing them. This was reinforced in the short-lived children's version of the app—Snapkidz—which disabled sharing functions but retained these image modification features.[76] By allowing children to craft and create visual forms on their screen frames, Snapkidz and similar apps are the contemporary equivalent of twentieth-century crafting frames. In essence,

Snapkidz provided children with training in the techniques of networked image sharing that, once they were thirteen and able to download Snapchat, could be converted into immaterial network labor.

As the ubiquity of these apps attest, the millennial fears of photography specialists concerning the ontology of the digital image have done nothing to diminish the photograph's popularity as a social tool. The digital image's remarkable ability to be stretched, shrunk, stitched together from multiple images, strung into patterned groups of images through tags and feeds, and to serve as base for the application of other visual materials has made it the basic cultural unit of contemporary networked interaction. We employ our phones and tablets as frames within which we work on this vast array of material, putting it together in new ways. Underlying software functions much like the settings on a sewing machine, allowing us to organize and combine this material according to standardized procedures and patterns. Perhaps it is no surprise, then, that hints of the earlier skepticism around digital images remain, now focused on the potential frivolity of such networked image practices. As Martin Lister claims, "Characteristically [such images] exist in multiples: as strings, threads, sets, grids ... " that produce less attentive, more distracted viewers. He continues: "This fugitive and transient networked photograph and its restless viewer (or user) is more than an aesthetic form. It is part of a larger reconfiguration of experience and mediation of the world by information technologies."[77]

# 3   Piecework

We find ourselves bent over our mobile touchscreen devices at moments throughout the day. We hold them in our hands or prop them on our laps. The proximity of these devices to our bodies, and the relatively small, sometimes imperceptible finger gestures we make on their surfaces can convey dedication and intensity of focus to those looking on. Against the rhetoric of pleasure, ease, and leisure that has surrounded mobile devices since the rise of the iPhone, it can be startling how much this focus and these actions look like work. In a 2002 study of creative applications of digital technology, Ben Shneiderman asserts: "Information and communication technologies are most appreciated when users experience a sense of security, mastery, and accomplishment. Then these technologies enable users to relax, enjoy, and explore."[1] Writing only a few years before touchscreen phones and tablets would redefine the digital experience, Shneiderman nevertheless points to the sources of the tension that lies at the heart of the participatory, networked culture these devices have facilitated. In the design of their hardware and software, these devices permit an unprecedented sense of mastery for the user. Slight variations in the speed, direction, and force of our touch can produce very different results. This can engender a strong sense of enjoyment and relaxation. However, in our networked existences we often attain these at the price of a sense of accomplishment. Accomplishment eludes us because our networked presence requires continuous inputting and updating. Indeed, Wendy Chun argues that the process of constant updating is the defining condition of contemporary networked life.[2] We do not reach the end because, as is implied with interface features such as infinite scroll, there is no end to be reached in networked interactivity.

Chapter 1 considered the long-standing role textile culture has played in the development of digital technologies, including gendered and racial

divisions in labor and economic opportunities in the electronics industry. Chapter 2 explored the relationship between digital processes and
images and its contribution to networked interfaces. This chapter draws
on both subjects to interrogate the conditions of contemporary mobile
media—their devices, apps, networks, and physical environments, as well
as combinations among these—as they relate to theories of labor and social
formation. It argues that the structures of contemporary, participatory
media—particularly what has been called "social" media—foster the production of capital through uncompensated (or poorly compensated) yet
incentivized labor under the guise of leisure, entertainment, and sociabil-
ity. On the face of it, there is nothing extraordinary in such a claim. We all
know, for example, that the more than $300 billion market value of Face-
book rests almost entirely on the frequent participation of its many users
and the content they create.[3] What would Facebook—or any other social
media site, for that matter—be if its users stopped posting and sharing content? This chapter seeks to demonstrate that this imbalanced relationship
emerges in part out of participatory media's affinities with handicraft labor
and textile culture. Personal touchscreen device use resembles the actions,
strategies, and conditions of craft production, thereby casting such digital labor as "soft" labor that need not be compensated. Engaged in social
media participation, these actions enter into a process construed as leisure,
rather than the extraction of affective, immaterial labor. Like needlework,
this labor often takes place under circumstances and in contexts that would
be considered downtime or free time. Like textile assembly, much of this
activity entails taking extant material—in this case, data in the network
accessed through image-based interfaces—and binding it together in new,
flexible patterns and formations.[4]

It is important to realize that the ergonomics and self-management of
the activity of network assembly are honed not only through social media,
but also through many of the tasks and goals presented in mobile gaming, which bear comparison to needlecraft practices. Unlike console games,
game apps designed for mobile devices often center on tasks of matching,
coordinating, and connecting. These small, repetitive actions conform to
the constraints of handheld touchscreen interfaces and the everyday reali-
ties of their use in tight spaces and short intervals of time. The connective
aspects of their gameplay fold back into the parallel assembly work of social
media, sometimes in direct ways such as sharing scores or requesting more

game materials (e.g., "lives") from contacts, much as a seamstress might ask a neighbor for an additional spool of thread or a new needle. On occasion these games are embedded within social media platforms themselves.

This new affective, immaterial labor of mobile media use bears important affinities to earlier forms of labor typically gendered as "women's work" and often bearing additional racial and ethnic overtones. It intersects two important forms of such labor. First, as off-site and off-the-clock assembly of materials for corporations that will convert these assembled products into profit, this work mimics the structures of home-based industrial assembly known as "piecework" or "homework." In its precariousness and invisibility, piecework historically has been a labor form that exploits women who must care for children at home or are undocumented immigrants. Second, in its concentration around social interaction and networking, this labor draws from the tradition of informal, but ritualized group assembly of material objects, such as the quilting bee, that have coupled collective labor with affective community interaction in networks of friends and family. With their emphasis on material production, quilting and the quilting bee have been forms of collective action available to disenfranchised groups and people of color. It has been the social realm of those marginalized by the visual economy of the public sphere, which is why it has been so significant to women, and particularly to women of color, as a social and cultural practice (figure 3.1).[5] The ability of mobile social media and gaming to intersect these two historical and ideological trajectories has contributed to their remarkable power to extract labor for little or no financial compensation.[6]

## Pieced Networks

More and more interfaces are adaptable displays of material that may differ from device to device based on user histories and preferences, hardware and software specifications, and geographical location. Webpages and apps become flexible templates, therefore, within which underlying algorithms may rearrange, substitute, or replace material as necessary. This aspect of template design is also an important consideration for the user's physical interaction with the presentation, as when orientation rotation in rectangular handheld screens allows users to turn the device in their hands to alternate between vertical and horizontal display modes. While

**Figure 3.1**
Jennie Pettway and a small child with the quilter Jorena Pettway, Gee's Bend, Alabama, 1937. Photograph by Arthur Rothstein, U.S. Farm Security Administration, Library of Congress photo LC-DIG-fsa-8b35946.

the following examples underscore the fragmented, modular, and piece-meal properties of networked interactions as dictated by hardware and software infrastructures, what is more significant in the current context is how user interactivity and production is similarly constructed around models of fragmentation and remediation. This is an essential feature of networked labor. In 2002, David Weinberger wrote that the Internet was "small pieces loosely joined." Speaking of the move away from printed books to screen displays, he explained: "What once was literally a tightly bound entity has been ripped into pieces and thrown in the air. … And most important, the Web is binding not just pages but us human beings in new ways. We are the true 'small pieces' of the Web, and we are loosely joining ourselves in ways that we're still inventing."[7]

With the rise of handheld touchscreen devices, network interfaces—from operating systems and available user settings to platform design, functionality, and outputs—have become increasingly aligned along an axis favoring frequent, short-term, diversified use. The interface with the network is designed to be forever ready to let you pick up where you left off, or let you switch between tasks. It rewards frequent returns through constant updates that solicit some sort of response. Social media platforms and game apps represent major components of this production and interactivity. Two essential mobile media practices to consider in this regard are tagging and casual gaming, both of which trade on techniques, strategies, proficiencies, and affective value generation common to needlecraft culture.

## Hashtags and Tagging

If the interconnectivity of social media could be reduced to a sign, it might be the cross-stitched character of the hashtag: #. The interface and network's quintessential visual indicator of their processes of linkage and assemblage, the hashtag appears in social media as the prefix to a word, a string of words, or an acronym, marking them as searchable metadata links. Selecting the tag "#handicraft" when it appears within a platform's interface, for example, sets off a command to retrieve and display as an ensemble all contributions to that system that have been marked as such (subject to certain restrictions, such as privacy settings). The hashtag is the stroke that connects accompanying material—for example, a text entry or a photo—to past and future contributions to the system that bear the same tag. In essence the hashtag functions like the stitch that, executed multiple times, binds all of this related material together. The quilted display of contemporary networked image culture is composed not only through complex algorithms of search engines and programs, therefore, but also through user-generated data aimed at that purpose.

The hash sign itself (also known as the "number" or "pound" sign in North American English) is of ambiguous origin and has long been used for multiple purposes in computing languages. It came into use in Internet Relay Chat (IRC) as a means of labeling topics and groups such that they would be visible across an entire network. Hashtag use as it exists in social media practices originated around 2007 on microblogging sites such as Jaiku, as a means of aggregating messages on a particular topic, and grew rapidly with the rise of Twitter.[8] In 2009 Twitter began hyperlinking hashtags to search results, allowing users to access all tweets bearing a specific hashtag by selecting a single instance of its use. The function quickly spread to other social media and content-sharing sites and apps.

The hashtag has become the primary means for the users and producers of images to link an image to others along different and multiple patterns. In most image-sharing applications, tapping on a hashtag produces a grid of images all sharing the common thread of that hashtag, thus rendering a patchwork of images that ostensibly bear a common semiotic relationship. This may be more or less evident visually, whether through the action depicted, the form, or even the colors or patterns visible across the images. Significantly, the cross-hatching of the hashtag can be said to visually resemble the material lineage of the operation to which it refers. In other words, the cross-hatching of the hashtag recuperates the cross-hatching of needlework. In this way, the hashtag moves from an arbitrarily symbolic relationship to what it represents to a more iconic depiction that visually resembles its referent. This correlates with the symbol's name, which may reference hashing as the action of fragmenting or bringing together scraps (especially in cooking).

Hashtags are the social media successor of hypertext (or hyperlinks), an elementary component of the Worldwide Web from its beginnings. As a link on a webpage—initially a word in bold or underlined, but now often an image or symbol—hypertext signals a stable, "permanent" connection to material elsewhere on the web. Selecting the word or image accesses the connected material. Theodor H. Nelson, who coined the term "hypertext," describes it as "text that branches and allows choices to the reader, best read at an interactive screen. As popularly conceived, this is a series of text chunks connected by links."[9] George Landow describes hypertext as "text composed of blocks of text … and the electronic links that join them."[10] Hypertext as the stitches in a patchwork is the organizing idea of Shelley Jackson's 1995 work of electronic literature, *Patchwork Girl*. Jackson's tale of a female counterpart to the Frankenstein monster is structured through hyperlinks. "*Patchwork Girl* is an intermingling of traditional women's arts, such as sewing and quilting, with high-tech, cyborgian, queer performances," writes Jenny Sundén.[11] Jackson explicitly associates the stitch, the scar, and the link as they shape her title character's physical construction as a gender representation. But it is important to point out that, rather than producing a patchwork, hypertext functions as a sequential link that allows passage from one piece of material or information to another. To an extent, each new frame of information will replace the previous one. We may become conscious of this chain when we follow a series of links only to find ourselves both facing material we had not initially intended to access and unable to recall how we got there. The hyperlink has represented, therefore, the endless possibilities and paths represented by the branches mentioned by Nelson.

Contrary to this sense of infinitude playing out one step at a time, new modes of networked engagement such as the hashtag retain the link as a constitutive element of the media experience while nevertheless opening up its potential to assemblage in smaller patterns of production that may not represent vastness so much as affective links binding material together. Rather than step-by-step linearity, the aesthetics of the hashtag relies on the selection of a single link that opens onto an assemblage of material—tweets, photos, or something else—presented as an integrated whole, but produced through the process of attaching its multiple pieces one by one to this whole through the dispersed—one could say crowdsourced—labor of many. In networked systems of image production and circulation, the hashtag therefore represents the added labor of building collections of material out of small fragments through the meticulous work of labeling each fragment for integration into the whole. This work is not the sweatless labor of the algorithm, but the toil of the user. Creating material, such as photos, and uploading them to the network, is only one aspect of contemporary digital practice. More often, networked practices entail tying together extant material through deliberate and incidental actions online. Popular hashtags such as #like4like and #follow4follow point to this labor. Intended to elicit actions from the viewer, liking or following the material or accounts bearing these hashtags makes visible the regular labor that composes networked experience, and touch interfaces have physicalized this labor of connecting at the surface of the screen.

## Matching and Threading

Touchscreen mobile devices have led to an app-driven proliferation of "casual" games, fundamentally changing our physical and psychological relationship with these devices. Casual games are distinguished from other digital games by their elementary rules, simple visuals, and modest goals suitable to short intervals of use. Like the production and maintenance of hashtags, casual gaming often involves simple tasks of identification and association. These characteristics are in strong opposition to "hardcore" digital games, such as console-based games and MOORPGs, which often require extended commitments of time to understand narratives, master move combinations, and build long-term strategies for survival. As one woman gamer described casual games made for mobile devices: "Say you're a mom trying to cook dinner, take care of chores, run errands, etc. If you're going to sneak in a little play-time, you need a game you can save often and pause at any moment. You need to be able to pick it up, play for 20 minutes or so, and run out the door—as opposed to a game which requires several hours of focused play to progress."[12]

Built to be played between or during other activities, casual games "address the issues of time competition observed in domestication studies and the gendering of leisure time in the home," according to Alison Harvey, by avoiding the time-intensive narrative modes of console games (especially before the introduction of Wii) that "necessitate the kind of leisure time that is not typically available to female or older game players."[13] Aubrey Anable explains that "what makes a game 'casual' is that it functions in the ambiguous time and space between the myriad tasks we do on digital devices; between work and domestic obligations; between solitary play and social gaming; and between attention and distraction."[14] While video and computer game use has historically been dominated by boys and young men, mobile media gaming has been particularly successful with women, who now outnumber teenage males in the U.S. game market and make up more than half of all game users in the United Kingdom.[15] With this trend, casual games have been derided as unsophisticated and frivolous in comparison to the sporting, shooter, and survival genres within hardcore gaming culture. "Linked as it is to other kinds of players (women, girls, people of colour, and older people), casual gaming is consistently the object of derision within digital play paratexts such as game magazines, websites, and advertising, framing this type of play as inferior," notes Harvey.[16] This perception is reflected in mainstream media coverage of gaming. "How has such a seemingly mindless puzzle of blinking and beeping so thoroughly captivated so many?," a Bloomberg analyst disparagingly asked in 2014 about Candy Crush Saga, the enormously successful app produced by game company King. "The answer is simple. King's games attempt to provide 'a sense of achievement.' The puzzles are just challenging enough—and short enough—to provide a little dose of progress. Basically, the company has replicated the fulfilling feeling of loading the dishwasher or flossing your teeth."[17] Even within the scholarly discourse of game studies, Anable notes, casual games have been overlooked or dismissed, perhaps in part because they can be "affective systems that operate on various levels as mediations of 'women's work.'"[18]

Casual game culture has been dominated by matching and puzzle games that bear affinities to textile and needlecraft culture. Bejeweled is a significant early example. Originally released by PopCap Games in 2001 as a web browser–based game under the title Diamond Mine, the mobile app version—with a title that eschews masculine connotations of mining for feminine connotations of jewelry—was released in late 2007, several months after the introduction of the first iPhone. It was downloaded more than 150 million times in its first year. Bejeweled's success was followed by Candy Crush Saga, first released on Facebook by King in April 2012. Available shortly thereafter as a stand-alone app, it was downloaded more than 500 million times by the end of 2013.[19] Bejeweled and Candy Crush Saga are examples of the "Match 3" puzzle game. Thousands of these games have been created for iOS and Android devices, making them among the most popular and successful formats of contemporary mobile app–based gaming. Just as using hashtags builds links across similar or related materials, Match 3 games involve identifying and bringing together common objects or elements, frequently in sets of three. Their associated tactics, actions, and aesthetics trade on those regularly found in textile crafts and needlework. Bejeweled, Candy Crush Saga, and many other mobile matching games involve a grid of simple, colored elements—such as gemstones or lozenges—that the player organizes into patterns. Matches are confirmed by finger strokes that either align an object with corresponding objects or draw a line through them, as though threading these objects together (figure 3.2).[20] These structures and actions replicate processes of quilting, beading, and other textile and sewing crafts. Indeed, a common effect in these games, where elements appear to cascade down the game surface and settle into the chambers of the grid to then be linked through finger strokes, bears a striking resemblance to the matrix tray and pattern mold system of core plane weaving at IBM in the 1950s. As Gordon Calleja explains, these

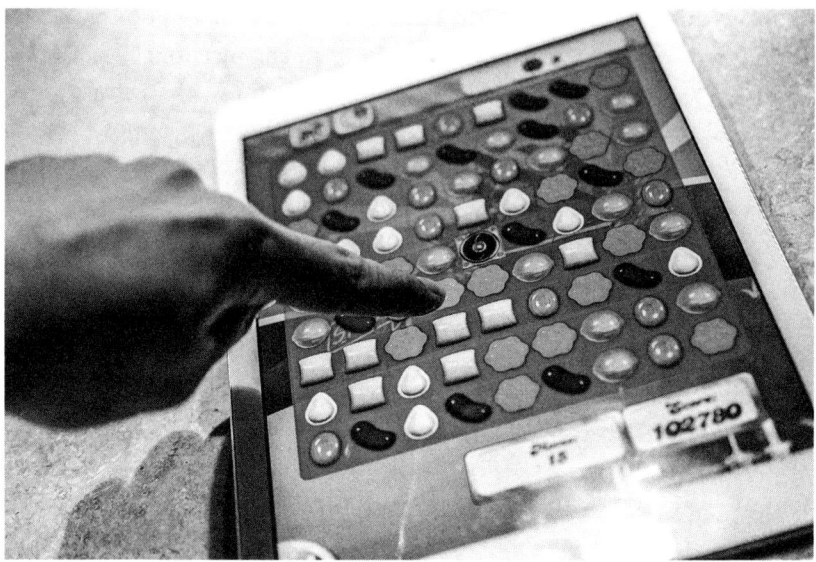

**Figure 3.2**
Playing Candy Crush Saga on an iPad, 2013. Photo by m01229.

games are not about projecting one's self into a game space, but instead entail manipulating objects on the surface within a bounded "totality on one screen."[21] In other words, the shape of the handheld device strongly determines the organization of the objects in its frame, much like the handloom or threading frame.

The gameplay of Dots, a matching game from 2013, bears an even stronger relationship to needlecraft aesthetics and practices while reflecting the wider relationship of these to mobile touchscreen interface design. Created by Patrick Moberg for Betaworks and developed by Playdots, Dots rose to number two among most-downloaded free iPhone apps in the weeks following its launch and had twenty million downloads in its first year.[22] The Dots interface is a minimalist, flat $6 \times 6$ grid of small circles in four colors. The goal is to string together circles of the same color, moving horizontally and vertically through the grid in a single finger stroke (figure 3.3). At the end of the stroke, the threaded circles disappear from the field and additional circles descend from the top of the grid, repeating the bead tray metaphor. Tracing squares or rectangles from circles of a single color earns the most points in Dots, because it triggers the removal of all circles of that

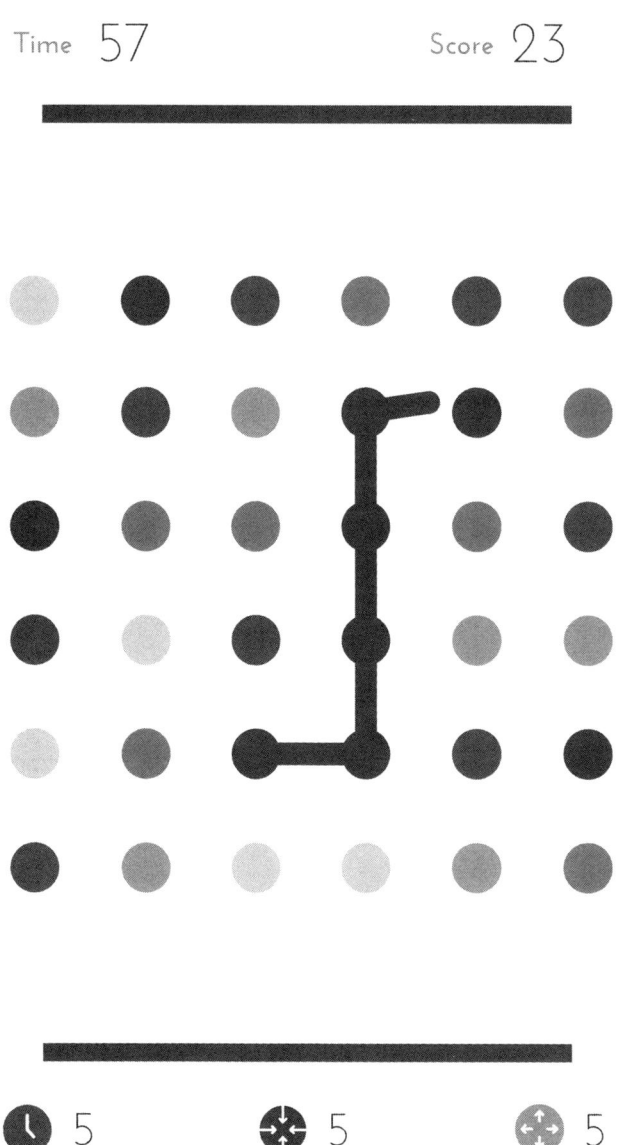

**Figure 3.3**
Dots interface. Image by Playdots, Inc.

color from the field. Moberg's decision to forego simulated objects for small monochrome circles produces a field that resembles preprinted embroidery or needlepoint patterns, which bear colored dots marking stitch points for thread of the corresponding color. The action of tracing the finger through as many circles of one color as possible reinforces this similarity, since the color of the connecting line produced on-screen matches the color of the selected dots.

Although Moberg has not described Dots in terms of textile practices, it is worth noting that the game derives from his work on Tapestry, a tap-based short story app he had developed before joining Betaworks. Experimenting with different interface designs and interactive possibilities, Moberg claims he focused on "what the iPhone does well, beautiful images, and smooth gestures."[23] Words therefore became colored circles, and the linearity of syntax and sentence structure became threadlike patterns across the device surface. "There is something hypnotic in the core mechanic of connecting the dots," Moberg explains. "Having a really tactile mechanic and building the game around that seems to be a good way to approach casual game design."[24]

The needlecraft labor simulated in these games is rewarded by accumulated points redeemable for additional plays or shortcuts. These can also be bought through in-app purchases that cost a few cents or dollars. But perhaps the greater commercial value of these games lies in their potential to fold back into the social network production described earlier. These games are designed to be shared easily through social media, to allow players to demonstrate their progress and post scores as well as rely on their social network for the additional aid of donated plays or game materials. Dots is not accompanied by in-app tutorials or instructions, for example, but was designed to allow easy sharing on Twitter and Facebook specifically to encourage circulation and conversation through these social media. "I think it's just a more meaningful experience," Moberg explains, "when you discover it on your own or from a friend as opposed to just like a computer telling you what to do."[25] The game and the labor it entails become the focus—and the by-product—of social interaction. The results of the pattern work produced, including screen grabs of particular moves or configurations of elements, double as a visual representation of the networked patterns the player likely already produces through interactions across multiple social media accounts.

## Quilting Affect

If hashtags, threading fingerstrokes, and other connective gestures in mobile media use can visually and materially suggest elements of needlecraft—particularly small-scale, home hand-frame production—the manner by which these are absorbed into a larger network of visual and social production reflects their deeper origins in the relationship between affective value and gendered labor. The goals and methods of social media regularly interpenetrate as the building and maintenance of social circles is produced through the collective generation and assembly of digital material, be it puzzle game achievements or vacation selfies. Social media users build their personal network by contributing these discrete bits of data for further assembly and reassembly as other users create additional patterns from the material not only through tagging, but also by liking, sharing, favoriting, pinning, tweeting, retweeting, or taking similar networked actions. This bonds the superficially formal, aesthetic aspects of the interface to the logic of informal, collective, crowdsourced labor that underlies distributed networks. As Matthew Fuller and Andrew Goffey explain, such a coupling of interactivity and production "is not so much an authoritarian or disciplinary means of making the brains run on time as a means of finding new grammars and techniques by which new things can collectively be generated and older things can painlessly be transducted into new contexts."[26] In the aesthetics of their interface and the ideology of their existence—joining pieced production and collective labor to ideas of affect, self-worth, and community identity—social media mirror many aspects of collective textile production and assemblage of the past two hundred years (roughly the span from Charles Babbage to Mark Zuckerberg), particularly those associated with quilting. Social media bear important parallels to the quilting bee, or social group creation for and through networked patterns. This emerges in social media's marketing rhetoric of "groups" and "communities," rather than the more formal language of "organizations" or "networks."[27]

Given the amount of material and labor required to construct them, quilts have always been prone to collective collaboration and bear strong social and affective connotations. Writing about the history of American quilts, Robert Shaw explains that, because quilts were usually given away or shared, they "connect people to one another as few objects can, invoking a sense of stability and continuity, a solidarity among those who have made them and all who have received and cherished them through the years."[28] Social media participation typically involves collecting existing digital media material or producing your own and assembling and sharing it through a network of social contacts. When used primarily as a personal, rather than professional, communication form, these flexible networks are typically composed of friends, family, and acquaintances. Craft communities can be a similar mix of people, and their purpose is often not only the production of objects but also the reification of social relations through those objects. In this regard, pieced quilting remains the quintessential example. Unlike whole cloth and appliqué quilting, pieced quilting involves the stitching together of individual swatches of textiles to create a larger fabric.[29] Once all of the fragments have been joined by seams, the entire assembled work is stitched to a backing stretched on a quilting frame. For larger works, this is often done on frames able to accommodate multiple people on each side, with each person responsible for stitching a specific area of the fabric, filler, and backing (figure 3.4).

**Figure 3.4**
Quilters seated around a quilting frame, United States, circa 1910. Image courtesy Rich701.

Pieced production is a global phenomenon and has existed for centuries—if not millennia—in most cultures.[30] This work has often been collective and female, shared by members of a family or community. In North America, European colonists and settlers created the quilting bee as a distinct cultural practice around the design, assembly, and execution of these fabrics. This collective labor model is particularly relevant when considering contemporary social media because both originate from an American sociocultural context yet they bear global implications and affinities. In nineteenth-century America, pieced quilting became a vital area of social production and control. As textile production was ceded to industrial mills that produced cheaper and cheaper cloth, quilting shifted from a practice based on salvaging valuable used material for new purposes to one that followed fashion, with the explosion of textile patterns available, while offering opportunities for social interaction in the sharing of materials and labor. In other words, the nineteenth-century quilting bee became more like contemporary social media production and interaction through sharing found materials (store-bought or gifted swatches in the first case, online photos and videos in the second) in a collective ritual of community tied closely to commodity culture. While a nineteenth-century quilter might compose her quilt of pieces of fabric gathered from friends to represent her overlapping and interconnected social circles, she might rely on templates provided by popular media for organizing and presenting that material. Periodicals such as *Godey's Lady's Book* and *Peterson's Magazine* regularly published quilt patterns for their readers, intersecting the deeply personal, homecraft aspect of quilting with standard patterns of presentation that could be repeated across communities (figure 3.5). Just as social media users rely on a platform's design and limited set of options—from overall site and account presentation to filters and effects (in the case of images)—nineteenth-century quilters were guided by explicit structures determined by mass media. "Throughout these years, as fashions in color, fabric, and pattern peaked and declined, the most important change was the transition of the quilt pattern from a folk art to a commercial industry," notes Barbara Brackman. This transition intersected personal and mass production in modern, industrialized communication networks and techniques.[31] One form of patchwork presentation would quickly replace another as tastes changed, much like one site or app might gain popularity over another, as with the

PATCHWORK.

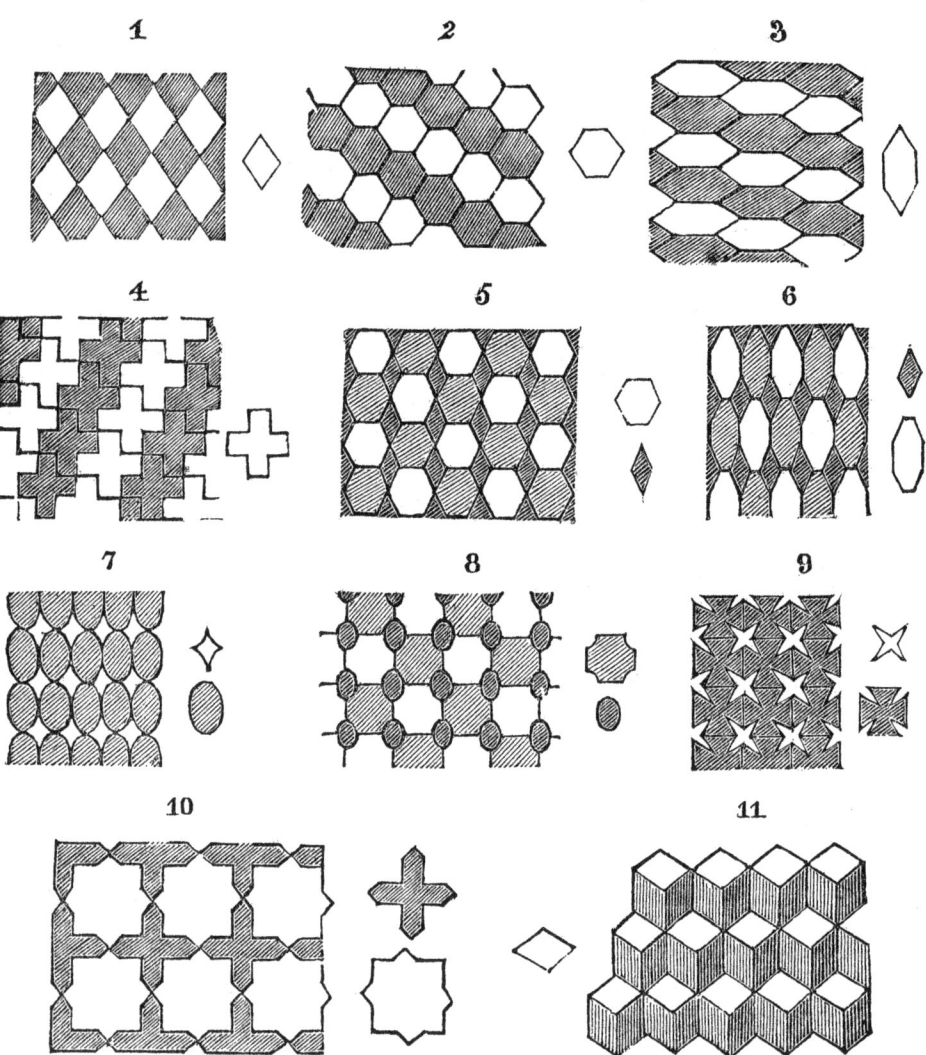

**Figure 3.5**
Patchwork patterns, published in *Godey's Lady's Book*, April 1850.

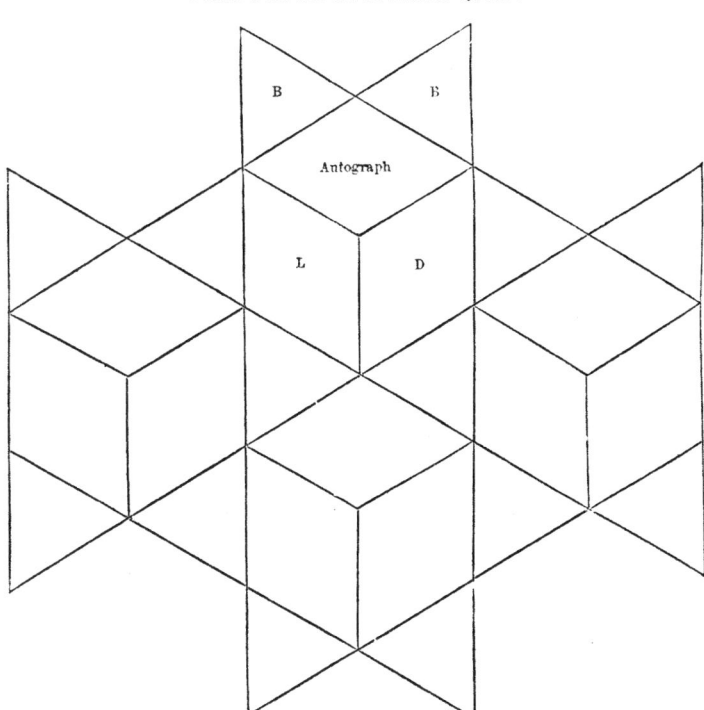

PLAN FOR AN AUTOGRAPH QUILT.

*Explanation of the Diagram.*—B B for black piece ; L for light ; D for dark.   The remaining blocks are finished in the same manner.

**Figure 3.6**
"Plan for an Autograph Quilt," published in *Godey's Lady's Book*, July 1864.

shift to Instagram from Flickr, Picasa, and other photosharing sites in the early 2010s.

A striking illustration of the quilt as social network appears in the July 1864 issue of *Godey's Lady's Book* (figure 3.6). "Plan for An Autograph Quilt" presents a quilt pattern designed to incorporate swatches of fabric bearing handwritten sentiments and signatures. Presumably the quiltmaker would solicit these autographs from her circle of friends and family, as was already common practice with autograph albums, and then stitch them into a pattern of light and dark material that offset each inscribed fragment as the top of a cube. The finished fabric would represent a network of social and affective relationships by incorporating indices of personal identity (the autographs) with other visual and material representations of social inter-action (for example, swatches of fabric procured through the everyday gift economy of sharing bought textiles and personal items of clothing). Like one's Facebook wall or photosharing feed today, this quilt would represent the components and connections of an individual's social community at that moment.

The extent of this assembly process and its close resemblance to contem-porary social media practices is reflected in the many popular stories about quilts and quilting that circulated in the nineteenth century. "Aunt Bina's Quilt," a story published in the *Youth's Companion* in 1898, is a good exam-ple. Written by Mrs. O. W. Scott, the narrative focuses on the making and handling of a quilt by a Massachusetts woman during the American Civil War (that is to say, at the same time the *Lady's Book* published its autograph quilt design). Aunt Bina Emerson assembles her quilt only from those calico pieces she has received from friends and townsfolk she likes and admires. "I won't have any pieces in it that call up anybody that's stingy or stuck-up or meddlesome or cruel … ," she tells her sister, "so when I'm sick it'll seem like a nice bright story."[32] Once she's finished the piecing, Emerson holds an exclusive quilting party "to which every woman came who was invited, for it was well understood by this time that goodness as well as gowns … was represented."[33] She eventually decides to donate this "love quilt," one of her most prized possessions, to the war effort. The quilt finds its way to a Washington, DC, military hospital, where by chance it is given to an injured soldier whose mother and sweetheart are among Emerson's clos-est quilting friends. The soldier recognizes his faraway loved ones in the assemblage. As the story continues: "He pulled the quilt nearer. It was made of many, many small triangles! 'Mother's dress,' he murmured, placing his finger upon a brown bit with a tiny white spray in it. 'Hetty!' and a wave of color rose to his pale face, as he caressed a triangle of pink. … Life was sweet after all."[34]

"Aunt Bina's Quilt" demonstrates the networked qualities of a quilt as a material and affective object in ways that find correspondence in contem-porary, networked digital culture. While Scott's story is a work of fiction, the situations it relates were indeed part of quilting culture and the value of quilts within society. As Cuesta Ray Benberry and Carol Pinney Crabb point out, "Almost every general interest periodical or quilt publication has printed stories about scrap quilts made of fabrics from garments worn by friends and relatives."[35] Sharing one's old or discarded clothing and linen with other quilters was not merely a means to producing aesthetic variety within the quilt, but also a way of reifying multiple, interconnected social relations in a narrative, visual object to be shared with others.[36] "The interchange of 'pieces' among neighbors furthered greatly, not only the artistic ends of quilt-making, but human friendliness and contact as well," wrote one quilting enthusiast early in the twentieth century.[37] Jack Bratich has suggested that quilting and similar collective textile practices amount to "peer-to-peer textiling."[38] Sharing such personal material was usually coupled with sharing the labor of assembly. A woman could measure her changing status within a particular social network by the presence and placement of her material in the quilts of others or the number of invitations she received to contribute her needlework to their completion. While the denouement of "Aunt Bina's Quilt" might be far-fetched, finished quilts were indeed shared with loved ones who would have a meaningful relationship with the people who had contributed their personal mementoes—dresses, handkerchiefs, bits of needlepoint or embroidery—to its overall construction. And this sharing often took place across considerable distances and despite great hardships.

## Homework

Historically, quilting extended across social strata—from the poor to the affluent—and was practiced by urban as well as rural women. Quilts were made out of necessity or as decorative objects. Artistic constructions with utilitarian applications, quilts also offered a departure from other household tasks, merging labor with leisure in the home. "Quilt making is a simple art and lends itself to many pleasant interludes—while listening to the radio, chatting with friends, or as a bit of 'pick up' work between household chores, for a quilt is made during a period of time and not at one sitting," wrote Marguerite Ickis in the 1950s.[39] The quilt represented the intersection of work and leisure, the merging of the social and familial, and women's necessity to multitask. Lucy Lippard's feminist reading of the quilt aptly explains this. She states:

The quilt is a diary of touch, reflecting uniformity and disjunction, the diversity within monotony of women's routines. The mixing and matching of fragments is the product of the interrupted life. Quilts also incorporate the grid, a staple of women's art in the early days of the feminist art movement, symbolizing, perhaps, the desire to salvage order from domestic and other distractions. What is popularly seen as "repetitive," "obsessive," and "compulsive" in woman's art is in fact a necessity for those whose time comes in small squares.[40]

When read with mobile social media practices in mind, Lippard's description is compelling, as it brings attention to the conditions of women's labor and demonstrates how the quilt is suited to that condition, becoming both its representation and its product. Participatory online media that rely on user creativity and labor to produce content to be inserted and circulated through their prescribed platform structures are similarly adapted to—and visualize—those "small squares" of time. The labor of relatively monotonous production is scattered across the day and night, in small intervals, between and even during other activities and tasks. Unlike quilting, however, this labor not only produces material to be shared within one's social network, but also generates quantitative value. For the user, this is measured in likes, favorites, shares, retweets, and similar reward structures. Direct financial reward is occasionally available, as in YouTube's profit-sharing mechanisms for content copyright owners, but this is the exception rather than the norm. For the proprietors of the app or website, however, value is indeed measured in profit, whether it is selling data, personal information, advertising space, or a combination of these. This turns social media interactivity and production away from the strictly personal and social uses that align it with the culture of quilting, toward models of gendered labor that historically have exploited the interrupted, distracted conditions of women's lives. Indeed, social media ingeniously entwine the social production of the bee with the economic production of piecework. This encompasses an expansion of user labor that, while often sedentary and monetized (if not recompensed), is similarly framed as leisure, mobility, and social engagement.

Our interactions with our mobile networked devices become "a diary of touch" for ourselves and our friends, as well as an economically valuable data history for the underlying platforms and networks attracting and facilitating that engagement. Quilting and contemporary networked culture share a reliance on mobile small-task production tied to affect and self-worth. The affective labor of networked culture—whether adding content to one's social media stream, responding to or tagging others' posts, or sharing hacks and amassing points in networked games—crosses the communal aspects of craft culture with exploitative digital labor conditions extending from the gendered industrial labor of homework. Most of us know, but conveniently ignore, that much of our online social activity is for the direct economic benefit of major tech and media corporations. Our Facebook accounts and all their content is not just ours, but also Facebook's. As Instagram's parent company, Facebook also stakes claim to the content of our Instagram accounts. Social media profit from the extent of the community networks we've been able to generate through our online assembly work as well as from the amount of time we spend on these sites. Metadata is sold or shared with marketers, advertisers, and partner social media sites. Advertising space is sold and targeted based on what we make, how we make it, and with whom.

In coining the term "liquid leisure," Tony Blackshaw underscores how leisure and work—both components of social media use today—have become ambivalent and enmeshed concepts. Whereas in the past leisure has been understood as involving contexts and activities that lie outside workplace labor, Blackshaw notes that leisure and labor now flow into one another. This derives not only from the mobility of work via digital networks, but also the business trend to adopt workplace models that encourage creativity and play among employees. "The upshot of these de-differentiating tendencies is that just as 'work' has become 'leisure-like', 'leisure' has become 'work-like,'" Blackshaw remarks.[41] However, for those who historically have been economically exploited, particularly women and people of color, that line has long been indistinguishable. The few moments of leisure in their day have not constituted clearly defined free time but rather moments unexpectedly carved out of the endless stream of waged and unwaged labor and tasks.

More and more of what might be called our leisure time is filled with mobile media interaction.[42] "Contemporary leisure and mass communications," David Rowe states, "display an unprecedented degree of mutual, institutional interdependence."[43] Giving the example of networked users' modification and dissemination of mass media images, Rowe identifies the increasingly powerful intersection between capitalized media and "cottage" media production and consumption. He predicts that "leisure time, for the foreseeable future, will be devoted largely to, or deeply implicated in, the reception and relay of communicative texts."[44]

Like our mobile media use, textile crafts historically have often lain at the blurred intersection of leisure and work.[45] As women's labor historians have shown, industrialization introduced the ideology of "separate spheres" in family labor relations, where men typically worked for a wage outside the home, and the status of women working in the home changed as their labor became less clearly defined or recognized. "Housework was losing status," Susan Strasser explains of post–Civil War America. "The independent home-based work of the other half began to look like something other than work."[46] This change of status, Ruth Schwartz Cowan argues, was produced in part by the entrance of the equipment and products of the factory into domestic spaces. Industrialization reduced or eliminated the work of men in the home, while it either bypassed or augmented women's work there. "If the advent of manufactured cloth eliminated the need for women to spin … it did not in the least affect the need for them to sew," Cowan contends. The greater availability of inexpensive, machine-made cloth increased home clothing production, and the introduction of sewing machines only increased home production expectations for women. "The advent of the sewing machine eliminated the need to hire seamstresses but not the hours spent by the housewife herself."[47] The ambiguous status of home-based labor generated by the Industrial Revolution has persisted, and the rise of portable, networked devices in the home and other non-office situations for both paid and unpaid labor that is usually immaterial contributes to perceptions of online life and social media use as distracting and unproductive pursuits. Indeed, in her research on the pleasures and intimacies of tele-networked employment, Melissa Gregg points out that women's household labor can make online job-related tasks a more satisfying exercise. "Women are prepared to wait until the cooking and cleaning are done, and the rest of the house is asleep, to have time to work alone," Gregg explains. "Having time alone with one's paid work can even become a form of solace from other, dubiously recognized, labors."[48]

In his work on labor and entropy, David Staples identifies women's work as historically "zero-work" or "work outside waged labor, or value production beyond value, which capital has continuously sought to colonize, discipline, and subsume in the sense of the old (but certainly not passé) imperialist and patriarchal orders." For Staples, "women's work has long invested … value production not only with energy but with affect and subjectivity too." Zero-work continues today most notably in the "turbulent expansiveness of global production and value extraction."[49]

The immaterial labor at the heart of much personal online activity effectively functions as zero-work. Maurizio Lazzarato explains that immaterial labor involves the production of cultural content through contemporary computing skills, on the one hand, and "a series of activities that are not normally recognized as 'work' related to art, fashion, taste, and public opinion, on the other."[50] Paul Taylor and Jan Harris claim that such labor operates at two levels within digital culture. First, it adds an "informatic dimension" to existing labor practices. Second, it extends labor into areas that previously were considered "leisure, escape, or simply outside of the market."[51] In their work on coding as a creative practice, Geoff Cox and Alex McLean echo Taylor and Harris's belief. "Labor is becoming ever more informational and communicative," they observe, "leading to a situation in which all activities seem to have been turned into production."[52]

In the contemporary mobile media context, much of this immaterial zero-work is performed through the physical labor of selecting and arranging visual data on touch-sensitive screens. The value of this information is based in part on the patterns of this handiwork. The manner and sequence by which bits of data have been connected can be mined for significant relationships and larger patterns. Each of us may be working on one small segment of this much bigger piece. Or rather, we may be working on multiple small segments, never fully aware of where these lie within the whole.

This piecework extends across the spaces and contexts of our everyday lives—the bedroom, the bathroom, the dining table, the living room sofa, the bus or subway ride, the coffee and lunch break at work. "Given the tremendous mobility and shifting both within globalization as well as on the microlevel of everyday life," Staples explains, "value is now being produced more or less everywhere and all of the time."[53] Such neoliberal corporate access to our time, our attention, and our labor diverges drastically from the historical standards of much organized workplace production, from factory wage labor to the office. As Tiziana Terranova explains, "The expansion of the Internet has given ideological and material support to contemporary trends toward increased flexibility of the workforce, continuous reskilling, freelance work, and the diffusion of practices such as 'supplementing' (bringing supplementary work home from the conventional office)."[54] Although some aspects of contemporary conditions may be unprecedented, there is also significant overlap with the culture of homework that historically has crossed textile—and even electronic—assembly with the unpaid domestic labor of women.

Indeed, home-based assembly work has been a key engine of economic globalization.[55] As Eileen Boris states, while homework preceded industrialization in Europe, it accompanied the growth of industrial manufacturing in the United States. She continues: "Such labor was gendered from the start, relying upon the sexual division of labor within households and between the household and the larger community."[56] Homework applied to a range of industries and products, but apparel—from shoes, to hats, to gloves, and even costume jewelry—comprised a large part of production. With the invention of the sewing machine and its successful sale to thousands of women in the United States on installment plans, "most of the work for the sewing establishments was done by women at home," claims Strasser.[57]

Homework has generally been task-oriented—rather than clock-oriented—labor. As Strasser states, "Task-orientation was the essence of piecework done for manufacturers as well as of housework. [Women] could plan their own work and suspend a task to care for a crying baby or stir the soup. While their husbands increasingly worked with others in factories, women continued to work alone or in very small groups."[58] Homeworkers past and present usually perform only a small part of the total labor required in making an object. They are paid by the piece, and "piece prices were systematically lower than those paid in the factory, often for the same work," states Cynthia Daniels.[59] Homework was curbed in the West in the twentieth century, but even when prohibited by law (as in the United States in the 1930s), it has never been fully eradicated. Taking place in the home, with a dispersed, little-documented, and even informal workforce, it has been difficult to track. Though much U.S.-based homework was in the garment industry, electronics companies sometimes depended on it as well.[60] For example, in upstate New York in the 1970s, "hundreds of rural, nonethnic, nonimmigrant, working- and middle-class women assembled basic components such as transformers, coils, and circuit boards in their homes for both national and multinational electronics firms," asserts Jamie Dangler.[61] "Research on homeworking in Silicon Valley dispelled the myth that homework is a unique problem of the needlework industries and thus an antiquarian survival of a previous era," according to a related study of electronics industry homework in New York in the 1980s.[62] "[Homework] exists today for the same reasons that supported it 50 and even 150 years ago," wrote one sociologist during this period. "For workers—mainly women with young children—it is a way of combining household commitments with paid employment. For employers it is a way of reducing overhead and keeping wages down."[63] Let us not forget that today these devices or their components are often assembled by women working in assembly "houses" in Asia, as explained in chapter 1, rather than in more formal and better monitored factories.

## Network Weaving

Feminists studying digital networks and culture in the late twentieth century anticipated some of the social and economic changes produced by mobile media. None other than Donna Haraway in her 1985 "A Cyborg Manifesto"—one of the most important documents of cyberfeminism and a founding text of digital studies—identifies the interdependent relationships of gender, labor, and new digital networks based on the homework model.[64] Haraway argues that communicative technologies are central to the feminization of labor through the growth of a homework economy. "Work is being redefined as both literally female and feminized, whether performed by women or men," Haraway states. "To be feminized means to be made extremely vulnerable; able to be disassembled, reassembled, exploited as a reserve labour force; seen less as workers than as servers; subjected to time arrangements on and off the paid job that make a mockery of a limited work day."[65] Although this important text has impacted critical technology studies in myriad ways, it has not yet been applied to the circumstances of today's portable, networked media. While Haraway saw the disintegration of existing social structures in labor's passage through digital networks, it is important to note that she did not anticipate the extent to which labor and leisure would interpenetrate. Indeed, the disassembling and reassembling that she identifies in the labor force has come to define not only this interpenetration of work and play, but also the immaterial labor of making and sharing networked content that both of these modes of experience entail. Haraway draws an important analogy between networking and weaving in suggesting a means of resisting these patterns. "'Networking' is both a feminist practice and a multinational corporate strategy—weaving is for oppositional cyborgs," she asserts.[66] With the networked practices described in this chapter, however, aspects of textile production—especially when related to the conditions of the homework model—must be understood as equally susceptible to exploitation as they are to liberation.

In the late 1980s, artist Martha Rosler also rightly predicted that the development of the digital image and digital processes—considered in detail in chapter 2—would have a profound effect on labor practices and worker rights. Speaking of the graphics industry, she explained:

> Computerization reduces the number of technologies involved in production and allows the work force to be dispersed, with the work done in the artists' own homes—which might be in Asia. This reversion to 'home work' (not in the school sense but as the term has been used in sweated industries like garment production) fragments the labor force, making not only conversation but solidarity close to impossible, producing a more docile group of piece workers.[67]

Today sites including Guru, Upwork, and Witmark provide freelance contract bidding for such work on a global scale. What Rosler did not anticipate, however, is how social media would make online visual production a socially oriented practice, deeply rooted in ideas of conversation and exchange, in a way that the attendant content creation for the commercial sites where these exchanges take place passes as something other than labor.

Rosler was writing at the moment when homework began to intersect with Internet-based telework, as white-collar employees made use of the growing intermeshing of digital infrastructures and telecommunications networks to perform more of their office tasks—from email to conference calls and project collaboration—while at home or elsewhere. This was the moment when the blurring of labor and other activities found in homework extended into the everyday in more general ways. Many more who did not formally or contractually work from home nevertheless began taking on certain employment responsibilities at home, in what might otherwise have been leisure time. In many ways, this set the precedent for what would emerge in the twenty-first century as Web 2.0. This shift in the Internet gave emphasis to the networked production of connections and content by users and consumers. Accessing the Internet meant building the Internet: uploading photos, sharing music, and writing in threads, chat rooms, or blogs.

A study of the social impact of teleworking—or telecommuting—at the start of the century noted the following: "For those who work at home, having increased access to technologies like the Web and e-mail at home may result in less time spent with family, one's community, or on leisure or household activities. Depending on one's perspective, this may be perceived as encouraging tendencies of workaholism or making it possible for people to spend more time doing activities they enjoy, which also happen to be work related."[68] As one woman teleworker explained, "'The drawback [of working at home] is you have to discipline yourself to ignore the pile of stuff you left undone in the bedroom that you really wanted to get to when it's time to get to work. … I'm a crafts kind of a person, I like to do projects. … Sometimes those things … *call* me to them.'"[69]

In many ways, social media have absorbed this project-oriented craft tendency. Social media are, after all, about building something: one's profile, one's network, one's value. It is a never-ending project. In combination with the aspects of homework and leisure noted here, social media not only contribute to, but in large part define, what Ned Rossiter has described as "organized networks." Rossiter notes, "There is a prevailing consensus that experiences of sharing, feedback, flexibility, and friendship are primary to the culture of networks." Although these communication networks suggest openness, he warns, they "are frequently not only not open, they also elide hierarchical operations that enable networks to become organized. Let us not forget that flexibility is also the operative mode of post-Fordian labour."[70] Jan van Dijk has gone so far as to identify social media as "a perfect illustration of the network society" in part because these media have "clear social orientation and sometimes entail altruism in sharing things."[71]

Amid this "altruistic" sharing, Terranova (like Rossiter) sees the encroachment of unpaid online labor into the everyday as a key component—perhaps the essential component—of contemporary networks and the functioning of networked society. She explains in her 2004 book, *Network Culture: Politics for the Information Age*, that digital media have produced the "social factory," extending outward from their networks into all aspects of social relations and the economy. "The Internet is animated by cultural and technical labour through and through, a continuous production of value which is immanent in the flows of the network society at large," Terranova

claims.[72] Working within an autonomist Marxist framework, she emphasizes the shift from the material production of the factory to immaterial,
online production that signals the continuity of consumer culture: "Free
labour is the moment where this knowledgeable consumption of culture is
translated into excess productive activities that are pleasurably embraced
and at the same time often shamelessly exploitative." Writing about the
situation at the turn of the century, Terranova cites America Online's volun-
teer chat administrators, "netslaves," and amateur web designers as "acting
out of a desire for affective and cultural production, which was none the
less [sic] real just because it was socially shaped."[73] The situation has only
intensified with the addition of billions of social media users and casual
contributors to online platforms.

In debates about online labor exploitation, Terranova detects a gendered discourse that distinguishes between coding and software design of
the sort found in the open-source movement, on the one hand, and the
content creation and management typically done on social media sites,
on the other. "This betrays the persistence of an attachment to masculine
understandings of labour within the digital economy," she explains. "Writing an operating system is still more worthy of attention than just chatting
for free for AOL."[74] Bratich has built on Terranova's theories in drawing
the link between such immaterial labor, gender, and the history of craft
culture. In pointing out "the informational and communicative practices
embedded in traditional craft culture," Bratich considers the same network
and craft links as tied to digital labor, which he calls "fabriculture."[75] Where
Terranova and others see rupture, Bratich challenges the autonomist reading of digital culture by delineating a continuation (he calls it a "thread")
of immaterial and affective labor running from modern society into digital
networks.

According to Bratich, craft culture's informational aspects align it with
immaterial digital value production. This emerges most strongly perhaps
in the similar social and affective production techniques and effects found
in the history of quilting and recent social media design and use. As demonstrated earlier, in both cases the material performance of this labor is
highly portable and adaptable to diverse situations. At its base, it also often
involves fingers interacting with small frames of material. In a nod to Terranova's theory of the "social factory," Bratich calls the predominance of
craft culture online (as with crafting groups and apps) an example of the

"social home," where the affective and communicative work of sewing circles and bees extends beyond the domestic sphere to the public sphere. Although Bratich's exploration of the social home is limited primarily to networked activity that explicitly addresses and promotes craft culture and production, in the current context it becomes a useful metaphor for the way ideologies and practices historically associated with textile-based crafts have informed not only the shape and functionality of hardware and software, but also the ways they are accessed and performed in everyday contexts. Like the online "social" actions of tagging, matching, liking, and sharing, touchscreen mobile media have an embodied, performative aspect that have brought the gendered legacy of craft and home labor into public view, if not public consciousness.

# 4    Domestic Disturbances

Mobility and corporeality are critical components of contemporary net-worked culture. Decreasing in size while increasing in processing speed and capacity, screen-based digital devices have become portable to the extent that we are encouraged—and expected—to carry one or more with us at all times. This can instill a more personalized and intimate relationship to the device, not only as an accessory, but as an object tailored to our body. Add to this the reactive surface of tactile screens and their increased options in orienting the frame and such devices truly entail bodily performances that can reshape the meanings of social spaces and the practices that unfold there.[1] These conditions differ significantly in at least two ways from inter-actions with earlier computing devices such as the terminal, the desktop PC, and even the laptop. First, those earlier forms often were confined to specific settings, such as the office, the laboratory, and the college dorm room. Before the rise of wireless data networks, even the laptop more or less adhered to this pattern, leaving such spaces only for business trips, client visits, and the like. Second, when the first popular wave of Internet migration took place—largely on PCs—in the late 1990s, the user's bodily performance was considered less explicitly tied to object, interface, and net-work. Indeed, entering the network through the keyboard and mouse was construed as a potential escape from the codes and constraints of every-day embodied living. Disembodied online existence arose as an attendant potential aspiration or threat of the network in its relation to user identity. Feminist scholars in particular saw networked digital culture as a respite or release from the gendered (and racialized) social conditions of physical reality. "If I have no body [online], what is my gender?" Mindy McAdams asked in 1996. "Is there a need for, or even an explanation for, gender in a place where our bodies are not?"[2]

Writing at the height of interest in virtual reality (VR) technologies in the 1990s, well before the introduction of smartphones and tablets into everyday culture, Ken Pimentel and Kevin Teixeira predicted—like many others—the imminent evacuation of hardware from the field of vision and sensory perception. "Virtual reality is where the computer disappears," they explained. "There's no little screen of symbols you must manipulate or type commands into to get the computer to do something. Instead, the computer retreats behind the scenes and becomes invisible, leaving you free to concentrate on tasks, ideas, problems, and communications."[3]

Neither the genderless disembodiment of the Internet nor the seemingly media-less immersion of VR has taken hold as a typical digital experience, however. Instead, digital interface in the everyday entails manipulating symbols on little screens, precisely as Pimentel and Teixeira described, often through touchscreen commands. In an essay on technological interaction and enjoyment published shortly before the success of mobile touchscreen devices, a group of human–computer interaction researchers asked: "People all have senses and a body with which we can respond to what our environment affords. Why, then, do human–product interaction designers not use these bodily skills more often and make electronic interaction more tangible?" Eschewing the possibilities of losing the body in the net or losing the machine in VR, they urged the following instead: "In order to design enjoyable products we must design for engagement on every level and the physicality of products should be restored. … In this way, not only functionality but beauty and fun in interaction are opened up."[4] The emphasis on device materiality, the bodily gestures of the user, and the potential pleasure generated by the interaction of these is a salient aspect of contemporary mobile media. With such media we often find new ways to adapt the device to our bodies and our bodies to the device. The device can become a material, visible extension of our bodies. Ideologies of VR—and related efforts to minimize interface—cast such involvement as distractions that occupy the body and make it difficult for us to be "free to concentrate on tasks." If, as Don Norman claims, "The real problem with the interface is that it is an interface. Interfaces get in the way. I don't want to focus my energies on an interface. I want to focus on the job," then one could argue that these devices and our attachment to them represent a sad regression in our relationship with communication technologies.[5]

The shift to mobile, handheld, tactile interfaces has important consequences for understandings of the social role and impact of networked culture. These devices and interfaces are structured to allow and encourage intermittent use that may be seen to occupy and restrain us. Although they may produce immersive experiences, they certainly need not, and they are easily manipulated while doing other things. This adaptability is embedded in hardware design and software functionality that facilitate and reward repetitive, tactile operations. Relationships between data are physically and digitally rendered through finger movements that connect them in visible and invisible patterns.

For the near-entirety of their history, however, media screens have been intended to be seen, not touched. In the phantasmagoria, magic lantern, and diorama shows of the eighteenth and nineteenth centuries, the screen was set before an audience much like the stage in a theater. With cinema's development at the turn of the twentieth century the theater model would become the norm, even influencing film projections in circumstances outside the movie theater, such as clubs, schools, and the home. The separation of screen and hand would remain even when the screen was within easy reach, as in the case of the television set or personal computer monitor. When the screen displays images or representational figures, this separation promotes the illusion of a continuity of space through the frame. Consider, for example, the annoyance that a stain or spot on the screen surface can cause for the viewer. Defective diodes in a screen's matrix may produce similar discomfort. These indications of the screen surface—of its presence "between" viewer and image—are reminders of the process of mediation. In principle, then, the separation of screen and hand contributes to an idea of realism that may hinge on either the idea of a space stretching beyond the space occupied by the viewer or the outright replacement of the viewer's space by the screen space. Television would be an example of the first—the small screen as a sort of portal opening onto another world. The cinema would be an example attempting the latter in its effort to minimize awareness of the theater space and emphasize instead the depicted screen space as the only world registering in the viewer's perception.[6] The intervention of the body within the image field—as when an object enters into the projection beam and throws a shadow on the screen—would be considered a nuisance in such a situation.

This has not been the case in all circumstances, however. Wanda Strauven, among others, has demonstrated how the idea of hands-free visual experiences has long been a guiding myth of visual media even as these media engage tactility in multiple, interactive ways that can play directly into the pleasure of image viewing.[7] For example, in the first years of cinema, when the projection of moving, photographic images was a source of fascination at fairgrounds and other sites of spectacle, touching the screen and exploring the mechanics behind the magic would be acceptable behavior. This was also occasionally true of earlier live projections from the camera obscura, according to Erkki Huhtamo.[8] But such examples remain the exception in modern screen history. Even as screens migrated into everyday living and working spaces in the second half of the twentieth century, an array of remotes, keyboards, mice, and joysticks allowed users to control them from a distance. Even when tethered to the screen device, these haptic interfaces rarely interfered with screen space or the user's line of vision.

In many instances today, particularly with smartphones and tablets, sight now vies with touch at the screen surface. The screen becomes the place for both the display and the control of the image. (Interestingly, the control of sound still is often accomplished by buttons along the frame edge, rather than through the screen.) This shift from the illusion of unimpeded vision has fundamentally affected our relationship to the image. For one, it undermines the privileged place of the gaze in screen media. With the onset of tactile screen interfaces as a norm—at least for mobile screens—the hand now regularly intervenes in the space between eye and image.[9] Any exploration of the relationship between image and vision in contemporary media therefore must acknowledge—and account for—this occurrence.

The steadfast veneration of the screen as a distanced, inviolate surface through most of its history places that object within a much longer lineage of practices and technologies emphasizing sight as the privileged sense in the sociopolitical sphere. Not only has the development of visual devices and media produced new ways of making, storing, and accessing images, but it has also codified physical and cultural conditions for seeing. It has contributed to a disciplining of vision that would identify and deepen social divisions according to race, class, gender, and other means of social differentiation.

This chapter considers the consequences of screen mobility and tactility for theorizing vision and the gaze as physical phenomena and ideological concepts. It suggests that the dynamics of mobile, tactile screen use threaten the dominance of visual regimes shaped by the modern history of the Western image. In particular, the negotiation of the visual through the haptic challenges hierarchies of the senses that have often been deployed in racial and gendered terms. Simply put, the intrusion of the hand troubles the distancing white European masculine gaze and bears consequences for gendered divisions in the material design and use of communication media. It places the screen within the register of the intimate, domestic object. This may have been the case with television, but the relative immobility of the television set meant that this domestic screen did not circulate within public spaces. The audience did not put it in their pocket and watch it on the bus or while waiting in line at the cash register. This mobile domesticity emerged, to an extent, with the first cell phones, as private conversations became overheard monologues in public settings. Yet, in contrast to touchscreen interactions, the phone conversation has little impact on the mechanisms of the gaze in the public sphere. When performed in public, the intimate tactile connection with a device can suggest withdrawal and distraction. In this transaction with the small screen, visual perception of the surrounding environment may be limited or secondary. Similar conditions have been observed in the execution of supposedly home-oriented exercises, such as sewing and knitting, when these have been performed in public spaces.

These changed conditions in the everyday use of collective spaces—where handheld devices make visible the connotations of private, domestic interactivity rooted in the tactile—may have contributed to recent efforts in interface design to restore the predominance of the outward-looking gaze through augmented or virtual reality. They may be behind Google cofounder Sergey Brin's reference to touchscreen mobile devices as "emasculating" in his promotion of Google Glass as an alternative communication device. The controversy surrounding Google's efforts to promote Glass as a means of greater engagement in public spaces and social contexts is an important example to be considered here. While Google marketed the Glass interface as nonintrusive and liberating, negative responses to Glass were partly based on its connotations of an intruding, controlling gaze. On the heels of Glass's failure, however, a range of eyewear has emerged to offer consumers varied means of creating expansive virtual fields of 360-degree, 3-D imagery built from image-stitching and other image-processing techniques. Whether Google's own low-tech Cardboard, which embeds the touchscreen phone in a paper viewfinder, Facebook's Oculus Rift, Samsung's Gear VR, Sony's PlayStation VR, or another headset-based system, these technologies attempt to reassert a sense of mastery over the visual field through immersive experiences that screen out surroundings with the illusion of an unimpeded, immediate, liberated gaze. The threat of intervening tactility that might disturb the effect is expunged through perhaps the last means available of ensuring touch's exclusion from the field: mounting the screen mere millimeters from the eye in a casing that leaves no room for roaming fingers.

**Constructing the Dominant View**

Throughout much of the history of Western thought, theories of consciousness and perception have posited the visual as predominantly masculine in its social and cultural functioning. Visual regimes such as the one-point, linear perspective developed during the Renaissance—or what Teresa de Lauretis has called "the system of the look"—have reinforced hegemonic masculinity within the social and political field.[10] These regimes deploy and sustain ideologies of gender through the physical organization of the visible environment and the establishment of social codes of looking. In this framing men see, while women are to be seen.

   In considering social order, power, and the senses, philosopher Jacques Rancière has identified an "aesthetic regime of politics" that first operates on the visual plane. Rancière states, "Politics revolves around what is seen and what can be said about it, around who has the ability to see and the talent to speak."[11] Here sight and speech become intertwined to form a chain of agency and action to be exercised in the collective spaces that sustain the public sphere and political discourse. In her research into the social role of the senses across Western history, Constance Classen explains that "women were the ones who cooked and cleaned and sewed at home while men used their eyes and ears outside in the world."[12] As Nicholas Mirzoeff has demonstrated, throughout modern history institutional understandings and social practices of the visual regularly have served to marginalize and oppress segments of society that included women, children, the colonized, and the enslaved. Such parts of the population were—and often continue to be—supervised and overseen by formal and informal instruments of social power, especially when in public, while being severely restricted in—or outright denied—their "right to look."[13]

Within the mind/matter—or mind/body—split of Cartesian dualist epistemologies, vision is the sense closest to any supposed immediacy of the intellect. Other senses, particularly touch and taste, are designated as more deeply rooted in the body's material constraints, since perception through these senses requires direct bodily contact with other objects. Elizabeth Grosz theorizes that this dualism at the center of Western philosophy has fueled other binaries, including reason and passion, sense and sensibility, depth and surface, and male and female, establishing contrasts that potentially function in the opposition between virtual reality systems emphasizing sight and those mobile media practices that foreground tactility.[14]

In *The Phenomenology of Perception*, a work strongly emphasizing the role of the visual, Maurice Merleau-Ponty references the haptic in a few lines buried deep in his argument. He explains that the sense of touch constantly returns us to our body as the material prison of our subjectivity. "As the subject of touch, I cannot flatter myself that I am everywhere and nowhere," he observes. "I cannot forget in this case that it is through my body that I go to the world. It is not I who touch, it is my body."[15] In distinguishing the "body" as site of touch from the "I" of subjectivity, the haptic here reduces the being to an object while vision (the "eye" behind the "I," at least in English) drives the myth of a liberating and totalizing "everywhere and nowhere."[16] As we shall note later in this chapter, the myths (and marketing) of virtual reality rest on such an everywhere-and-nowhere as a technological possibility. That this mastery of the visual is at once powerful and illusory is essential to the relation between vision and tactility in contemporary mobile interfaces. If the historical dominance of visuality as a regulating force in the public sphere rests on a foundational illusion of a seamless totality of both view and consciousness, then a turn toward visual interfaces highly constructed through immersive visual illusions seems a "natural" means of recuperating that supremacy of the gaze. The visual device becomes a means of sustaining the myth, even as it is being undermined every day by tactile interactions with small screens that may be construed as fragmented and potentially disruptive practices diminishing the integrity of shared physical spaces, from the park to the classroom.

Maintaining this illusion may be ideologically important because, if vision is meant to be employed—at least metaphorically—in strategizing and theorizing, touch is meant merely for toiling. As film theorist Christian Metz states, "It is no accident that the main socially acceptable arts are based on the senses at a distance (such as vision and hearing), and that those which depend on the senses of contact are often regarded as 'minor' arts."[17] What then is the cultural significance of the new relationship between vision and touch as it plays out in personal screen use? Where do touch-activated and animated visual experiences fall on this spectrum of major and minor cultural production? How does this relate to the wider history of the social function of the senses? The history and meaning of the screen have long been shaped by these ideas.

In most instances, regulating the visual field—technologically or otherwise—has meant regulating women's place within it. This has been achieved historically through the evolution of law, architecture, and other means of organizing and enforcing the social production of space. Mirzoeff calls such efforts "modalities of visuality." For him, a modality has three steps: it classifies people within a population, it separates them according to these classifications as a process of social organization, and it "makes this separated classification seem right and hence aesthetic."[18] Recent work in gender and identity studies has contended with this process and how it has played out historically in defining gender, race, ethnicity, and other categories of classification and difference.[19] The media and their use of the screen have been important instruments in these processes of visuality. This was argued nearly a half-century ago in feminist and apparatus film theory by Laura Mulvey, Mary Ann Doane, and Stephen Heath and Teresa de Lauretis, among others.[20] Their theories of the transmission of ideology through mechanized visual media rest on the alignment of the viewer and the lens—both metaphorically and physically—as a means of naturalizing the moving image and, therefore, naturalizing hierarchies of social difference conveyed within the image's content. The hegemony of the darkened movie theater remains the quintessential example for this theoretical stance. That a modification of this model to recognize other forms of media consumption might disrupt this "naturalness" was illustrated in 2016, when the head of major American movie theater chain AMC Theaters announced that the company might allow mobile device use during screenings. In an example of the negative (and perhaps hypocritical) perceptions that surround public mobile media use despite its popularity, two days of significant negative social media backlash provoked AMC to abandon the idea through a tweet: "NO TEXTING AT AMC. Won't happen. You spoke. We listened."[21]

While historically sight has been the sense most closely associated with masculinity and its dominance in the public context, the tactile or haptic has been deemed the realm of women and the intimate. A "woman's touch" becomes the natural opposite to a "man's vision." As Classen describes it, "Women are the forbidden taste, the mysterious smell, the dangerous touch. Men, by contrast, have been associated with reason, as opposed to the senses, or else with sight and hearing as the most 'rational' of the senses. The occultation of the sensory underpinnings of Western culture by the modern visual and rational world view may therefore be read as an occultation of certain feminine dimensions of that culture."[22]

Writing in the 1980s about the growing range and reach of visual technologies, Donna Haraway tied these developments to the concept of an unfettered male gaze inhering to power structures from capitalism to colonialism and militarism. She declared, "Vision in this technological feast becomes gluttony; all perspective gives way to infinitely mobile vision, which no longer seems just mythically about the god-trick of seeing everything from nowhere, but to have put the myth into ordinary practice."[23] Some might venture that the "gluttony of infinitely mobile vision" that Haraway describes has been perfected in networked, mobile screen culture as the reach of the gaze glows in the screen in the palm of nearly every hand. Yet our hands do not just hold the screen. They tap and stroke the screen's surface. The finger intervenes in the viewing window. In other words, use of a touchscreen requires occlusion, only working properly when our bodies block—or "screen out"—part of the image in the frame. It forces our flesh into the picture, reminding us of sight's inescapable bodily envelope.

Cyberfeminism of the 1990s held out the hope that society might move beyond gender and other categories of social differentiation and division via the networks that screens have brought into view. For theorists such as Allucquère Rosanne Stone, cyberspace offered an environment that freed people from the socialized body and the repressive structures of gender as these played out in the material objects and practices of the everyday.[24] Indeed, the shift that fueled this belief was the seemingly reduced role of the visual (and the spoken voice) in networked interactions. Cyberspace was thought and expression let loose from their bodily origins. As the user could not see or hear the people with whom she interacted in chat rooms, listservs, and the like, the Internet appeared to offer new opportunities for gender-bending and even gender dissolution. "The gendered body theoretically does not play an important role anymore, considering for example the work at a computer screen—nor does it in the world of the net where the decision for a gender is practically of free choice," Priska Gisler claimed in her 1997 article, "Does Gender Still Matter? Bodily Functions in Cyberspace: A Feminist Approach." Nevertheless, Gisler noted, "As long as power structures are important in the net, gender will play an important role."[25] Anita Greenhill similarly claimed that "cyberspace allows a rethinking of the mind/body dichotomy because the processes of identity formation and sociality move beyond simple ascriptions."[26] As this book has argued, however, such has not been the case. Like any social construction, gender can transcend bodily conditions even as one uses the body to affirm or negotiate its codes, making gender differences just as alive and well online as they are offline. Gender returns, as strong as ever, in networked culture and in the way it plays out in our daily lives.

## Point, Aim, Shoot

Twentieth-century computing did more than its share to contribute to the hegemony of the visual. In the evolution of screen-based interfaces from the mainframe to the laptop, a variety of devices and accessories were developed to permit distanced interaction between the user and screen space, thereby maintaining the long-standing separation already present between viewer and screen in other media. The light pen or gun, the mouse, and the joystick—all invented in the twentieth century—facilitate interaction while maintaining distance between body and screen surface. Each of these devices derives as well from a militaristic logic rooted in the actions of steering, targeting, and shooting. The joystick—a staple of video games to this day—originated as a steering device in airplanes at the start of the twentieth century. It is perhaps fitting—though chilling—that joysticks are still used today to steer aircraft in combat, but now these aircraft are drones, and several thousand miles may separate the "pilot" from the target. The technology behind the light pen was first patented and exploited for amusement arcade shooting games in the 1930s.[27] The "Ray-o-Lite Rifle Range," manufactured and distributed by the Seeburg Corporation from 1935, was an ammunition-free shooting gallery that relied on a light-projecting rifle "built to regulation standards" that hit moving targets equipped with photocells.[28] The first Ray-o-Lite Rifle Range targets were die-cut metal ducks, but Seeburg sought to expand appeal by offering players opportunities to shoot metal bears, mothers-in-law, and, during World War II, Adolf Hitler and "Japs." The technology would reemerge in the 1950s as the light gun in the Whirlwind computer at MIT (figure 4.1).[29] The photocell embedded in the gun would receive the light pulse emitted by the target on a glowing cathode-ray tube (CRT) screen.[30] Just as Whirlwind represents computing's hidden reliance on textile skills for fast, effective functioning, the project also represents a defining moment in digital computing's adaptation to a totalizing male gaze. From its raison d'être to its architecture and interface design, Whirlwind emphasized and expanded the role of vision in surveillance and subjugation, providing the visual with an underlying foundation of powerful, invisible calculation. Although the Whirlwind project began with the idea of building a flight simulator during World War II, as the war neared its end the project survived—indeed thrived—as the construction of what became the SAGE defense system of surveillance against high-altitude bomber attack. This modification moved Whirlwind away from the production of a confined environment that emphasized calculation and comparison to one emphasizing visibility by allowing the detection of flying objects and the depiction of their interception and destruction through interactive CRT screens.[31]

The CRT screen was the central link in the system. Representing real-time radar data of a patch of the North American skies delegated to a specific SAGE station, the glowing dot of an aircraft could be selected by the radar technician through the light gun. The gun had a pistol grip, and in its form and positioning (to say nothing of the military context within which it was used) it recalled a firearm.[32] "The decision as to which aircraft tracks are targets and which are interceptors is made by people and inserted manually into the devices by means of a light gun," a 1953 report explained.[33] Since intercontinental ballistic missiles made SAGE obsolete by the time it became fully operational in 1961, should it come as a surprise that military technicians turned these interactive guns toward other, more ludic targets drawn from masculine popular culture? SAGE maintenance personnel created at least two programs, known as "Girley 1" and "Girley

**Figure 4.1**
Demonstration of the Project Whirlwind light gun, 1952. Photograph used and reprinted with permission of The MITRE Corporation. © 2016. All other rights reserved.

2," that depicted women as animated sexual objects. Purportedly devised as diagnostic programs to ensure the system's proper functioning, these were run regularly at SAGE installations across North America in the 1950s and 1960s. Girley 1 recreated a 1956 *Esquire* calendar girl in a bodysuit as a static vector drawing on the CRT screen. Girley 2, also called "Hula Girl," was an animated, topless woman in a grass skirt.[34] If a technician aimed at Girley 2 and pulled the trigger as she swayed her hips, her skirt would fall off.[35] When SAGE was dismantled in the early 1980s, none of its light guns had ever targeted a Soviet bomber, but they had been aimed regularly at these images of women.[36]

The light gun is one of the few instances of a pointer entering into the field of vision in the twentieth-century computing environment. Another significant example is the light pen, based on the SAGE gun and developed in Sketchpad, Ivan Sutherland's 1963 graphical user interface (GUI) program.[37] Despite its references to drawing, Sketchpad sustained some of the aspects of distanced viewing already in place with SAGE. Like the light gun, the light pen did not touch the screen. "In its transition from gun into pen, the light pen retained its function as a targeting device," notes Branden Hookway.[38] Sutherland spoke instead of the pen as "seeing" and "aiming."[39] The codes of targeting and shooting were reinforced in the on-screen depiction of the pen's position as crosshairs.[40] Reminiscent of SAGE's Girleys, Sutherland's description of Sketchpad demonstrated its ability to produce "artistic drawings" through the examples of a winking woman—presumably acknowledging the gaze of the technician—and a woman traced from a photograph. The latter woman was depicted in two versions in Sutherland's dissertation: the original Sketchpad sketch and an alternate with different eyes and mouth that suggested the system's potential for modularity.[41] The pairing resonates with Rancière's claims concerning a polity based on vision and speech. "In refitting the features to the blank face," Sutherland remarked, "we discovered that, although the original girl was a sweet looking miss, an entirely different character appears if her mouth is made larger."[42] While these illustrations may seem obscure examples, it is important to note that Sketchpad is not merely one early GUI among others. As Lev Manovich states, "Sketchpad deeply influenced all subsequent work in computational media."[43] It is therefore worth noting that in Sutherland's seemingly anodyne exposition of his program's potential for producing animation he embarks on a performance of gender norms surrounding aesthetics, the senses, and the role of the gaze.

Unlike Sketchpad, in most other forms of GUI from the 1960s to early 2000s hand controls were placed outside the field of vision. From the mouse to the touchpad, the indexical marker of the hand's presence has been the cursor on the screen. Although in certain cases the cursor may convert to the image of a small hand (as when manipulating large photographs in some image-processing programs), for the most part it is represented as an arrow, which becomes the indicative equivalent of the *pointing* finger tied to sight, rather than the *touching* finger of material contact. This indexicality can be transferred to touchscreen mobile media devices through networked locative media. In essence, the device becomes an extension of the index finger, pointing from its screen to the surrounding environment. Nanna Verhoeff casts this potential of networked mobile devices as a "performative cartography." Describing the process of geotagging, for example, she explains that "geotagging allows the trace of the past—of the navigation it can retrieve—to become in the present the deictic index from which the traveler looks, so that—as a third temporality of the index—the image points to the future where the traveler can go: a destination index."[44] Pokémon Go, the extraordinarily successful animated geolocative game app released by Niantic in 2016, functions in this way. Players rely on their device's network connectivity and GPS capabilities to locate, identify, and catch virtual Pokémon characters in their area.[45] While such locative applications have become an important aspect of networked mobile media, whether for executing daily tasks or enjoying a game, much mobile device use does not require or invoke such situated, indexical interactivity. Most often our interface with, and experience of, the device change little whether we are in the street, on a train, or in bed. They are as much about immobility as mobility, as much about placenessness as place. Indeed, even Pokémon Go can be played while riding a city bus, with the user seated and staring at her screen, thereby negating the route finding and physical exertion commonly ascribed to this game.[46] These contradictory traits may contribute to lingering apprehensions about mobile media as disruptions and intrusions in the continuity of public spaces and collective encounters. While such apprehensions have certainly diminished with the increased penetration of mobile media into nearly all social contexts, they still rise to the surface from time to time, even among mobile media users, as we cede more and more of our social interaction to the network.

## Touch and Distraction

Mobile touchscreen practices disrupt aspects of the visual ideologies that play out both with other digital interfaces and in shared public spaces. They trouble ideas of unimpeded views and interactions, instead connoting a user's potential distraction or diminished awareness of the environment. Malcolm McCullough has explored the characteristics and consequences of such media use in his theorization of an "ambient commons" born of the increased intermingling of social spaces, bodies, and networked technologies. For McCullough, the emergence of this commons rests on the role of "ambient interface," which mediates relationships between people, places, and objects through network processes. Whereas earlier interface design was targeted at deliberate tasks and purposes, McCullough notes a millennial paradigm shift that "brought technology along into the existing world, in all its messy complexity," creating new types of embodied interaction as well as new understandings of what it means to be attentive or distracted, engaged or removed, in social contexts. "The interface arts address the play of attention," McCullough explains. "The challenge is to develop new cultural responses in the interface arts, as citizen sciences, and for the information commons."[47] The draw of habit and the fear of crisis, both heightened by the ceaseless activity of the network and the digital layers that bring them to our attention through the device, also play a role in this perception of public distraction, as Wendy Chun has shown. Pointing out that "habit itself is changing: it is increasingly understood as addiction," Chun explains that "imagined connections and edges—things that remain—are traces of habits. In terms of social networking sites, the strength of a friendship—its weight—is gauged by the frequency of certain actions. More strongly: *information is habit*."[48] Networked software cultivates "addiction" by soliciting our attention through regular, structured fluctuations in the properties of the interface—pings, beeps, changing light patterns and intensities, and vibration—in a stream of alerts, push notifications, updates, and the like. Reacting to such stimuli and attending to the potentially insignificant or irrelevant information they signal can imply a diminished commitment to, or interest in, the social value of one's immediate surroundings and the flux of activity within it.

Any concern about the social impact of mobile, networked media, especially in public spaces, may rest in part on the belief that they diminish the user's engagement with the dialogic, communicative potential of the surrounding environment by drawing her eyes to a flat screen. Touchscreen mobile media are literally "superficial" in the way they can collapse space for the user, diverting attention from surrounding people and objects. Conversely, physical spaces without handheld devices, or virtual spaces such as VR, convey a strong sense of continuous depth. Without a smartphone in hand to seek or capture our attention, we are apt to give the space and people around us more notice. At the very least, and this is key in my current analysis, we are likely to *appear* more engaged or attentive to those around us, regardless of where our thoughts may be. Keeping our eyes fixed on the device in our hands, therefore, can be not only a distraction but also an effective means of signaling our lack of desire for any interaction. While VR may draw a user's attention away from surrounding physical space in ways even more potent than a smartphone, a VR user's actions and seeming mobility within this digitally generated space nevertheless suggest an alternative space and sense of embodiment, rather than a retreat from—or rejection of—space and the body's place within it.

The heightened flatness of the handheld device experience is often by design. Apple's "human interaction guidelines" for the iOS system supporting its mobile devices, for example, has encouraged software designers to create interfaces that minimize visual depth and realism. "Reconsider visual indicators of physicality and realism. Bezels, gradients, and drop shadows sometimes lead to heavier UI elements that can overpower or compete with the content," Apple explains. It also urges reliance on translucency to give the sense of an intervening surface. "In iOS, a translucent element blurs only the content directly behind it—giving the impression of looking through rice paper—it doesn't blur the rest of the screen."[49]

Gaming is one area where the changing balance between sight and touch produced by mobile, handheld media becomes noticeable. The material conditions and interface aesthetics of touchscreens has led to the rise of games far less reliant on perspectival depth, aerial views, density of detail, and other aesthetic elements than typically has been the case with console games. As pointed out in chapter 3, games that thrive on touchscreen mobile devices often depict planar grids that diminish 3-D effects and entail finger gestures that emphasize the surface structure. These games reinforce the screen's superficiality by tightly binding the visual and the haptic. Finger controls are not forced centrifugally to the edges, as they are with other gaming systems that promote an unimpeded view by relying on joysticks or buttons along the frame edge. Touchscreen users must be aware of both the image and their finger movements, a situation that forces them to peer down at their phones or tablets, obviously at close range. With a gaze directed toward the lap or ground, rather than up and out at the world surrounding them, touchscreen users in public contexts may appear disconnected from the environment and any events taking place there.

Mobile touchscreen devices alter the visual hegemony of public space by bringing a newfound intimacy into play. Heidi Rae Cooley has identified this intimate aspect of mobile media in the device's relationship to the body. She describes the relationship in terms of "fit" between device and hand. These devices are tactile objects as much as visual objects. Writing about the smartphone, she describes the hand-object "fit" as activating "a relation of interface through which vision becomes and remains tactile."[50] She points out, for example, how it is common for a user to continue holding a smartphone in her hand even after she has finished interacting with the screen.

In his history of the mobile phone, Jon Agar notes that with the release of the iPhone the user's relationship to the screen object changed fundamentally. "The 'constant touch' of the iPhone … absorbs my attention and even when it doesn't I find that I unconsciously reach for the familiar smooth weight," he claims, supporting Cooley's contentions. "My fingers, eyes and mind are absorbed. And I am not alone—I have been in full train carriages where every passenger was communing with his or her smartphone. Each in a private bubble of constant touch."[51] Regarding such intimacy, Apple claims that "using gestures gives people a close personal connection to their devices."[52]

While other communication objects or devices, such as the walkie-talkie and the camera, previously have occupied everyday contexts, they have not connoted the same sense of intimacy with the body. For one, tactility with these earlier devices was limited to manipulating levers and pressing buttons and plungers, rather than tracing patterns across their surfaces with the finger. These devices also primarily engaged speaking and seeing, that is those bodily actions understood as more distancing—and privileged—in the public sphere (as per Rancière). While the touchscreen mobile device can perform the same or similar functions as these devices, its tactile interface emphasizes finger strokes on a frame, thereby bearing a closer resemblance to earlier devices of home-based craft and assembly practices than these communicative technologies linked to surveillance and the masculine gaze.

With these differences in mind, one can turn to the new relationship between body and screen object in mobile media use as identified by Ingrid Richardson. "Mobile devices antagonize any notion of a disembodied telepresence that is sometimes seen as endemic to screen-based media," Richardson claims.[53] She identifies a "mobile-specific *mediatrope*" built from the complex relationship between body and mobile screen object, particularly in public spaces. "The various postures and embodied actions particular to mobile phone use in public places, and the correlative dynamics of attention-inattention, are quite specific to the body-mobile relation, which has emerged throughout the last decade," she wrote in 2010. For Richardson, small mobile screens move away from the upright face-to-face interaction found with TV and cinema to produce instead a more complex range of relationships: "The often dedicated frontal orientation we have towards larger screens becomes compromised both by our own mobility, the size and resolution of the screen, and the interrupted nature of mobile phone use."[54]

Any claims concerning the supposed intrusiveness of the mobile touchscreen device in the public sphere are likely to rest on three factors: how it may seemingly remove the viewer from the activities around her, how it suggests an intimacy or privacy of action—working on something so small that others may not participate, and how it relies on a device that circulates freely between private and public spaces. Here the device interface replaces, corrupts, or at least complicates, the interface of public face-to-face encounters.[55] A 2015 Pew study of American mobile device etiquette reflects this perception. For most respondents the use of these devices was acceptable in situations of waiting or transit, but was considered unacceptable in situations understood to promote social interaction, such as eating at a restaurant or attending a meeting.[56] As Henrik Kaare Nielsen points out in his study of the Internet and the public sphere: "In face-to-face interaction we offer our bodily integrity as a pledge to our commitment and credibility and at the same time demonstrate a degree of trust and recognition of the counterpart that we would not be able to communicate virtually."[57] The preceding traits of public mobile media use suggest a disruption, evasion, and fragmentation of the integral sphere he describes. In reaction to this, many public settings have stated or implied restrictions on mobile media use.[58]

Interestingly, characteristics of mobile media use that may be considered detrimental to public social interaction also, on occasion, have made domestic labor activities such as sewing and knitting appear inappropriate in public settings. Jack Bratich and Heidi Brush note this in their study of craft and digital labor: "What causes such discomfort about knitting in public? One might put it this way: Knitting in public is *out of place*. ... Knitting in public turns the interiority of the domestic outward, exposing that which exists within enclosures, through invisibility and through unpaid labor: the production of home life."[59] The out-of-placeness of this activity derives from its ties to private, domestic spaces, as well as its insertion of the feminine into the public sphere in ways perhaps not conducive to the male gaze. As Stella Minahan and Julie Wolfram Cox explain, needlecraft groups that evolve from online digital interaction into a collective public practice may represent "women gathering together in a 'third place,' separate from home and work, for social activity." They add, "That this place may often be the local pub, traditionally a bastion of masculinity, is also noteworthy, for *Stitch'nBitch* may also be understood ... as a new protest movement using craft as a subversive vehicle for comment on gender as well as on the increasing commodification of society and technology."[60]

Bratich and Brush cite a *Boston Globe* advice columnist's claim that knitting in public is "terribly rude."[61] Online advice sites offer similar perspectives. "The age of the Internet and social media has given rise to a lot of fretting about the loss of manners and common courtesy in society. ... Isn't it nice to think, then, that even the oldest social activities are just as confusing? Because there certainly is no rule that tells us if knitting in public is considered rude or perfectly reasonable," reads a Howstuffworks column. "While the person doing it might consider it nothing but muscle memory, outsiders could easily take it as a signal that they're boring or uninteresting. ... If you're in a public space where knitting might imply a lack of attention—whether it's your intent or whether it actually is distracting you—hold off those needles."[62] A Miss Manners column similarly brings up mobile media use in tackling the topic. "Of all the multitaskers who could annoy you, Miss Manners would not have guessed that knitters would top the list. There is a centuries-long history of ladies quietly doing needlework while remaining alert to what was going on around them," the columnist tells an inquirer annoyed with knitters at religious ceremonies and cultural performances. "If they don't already ban texting, you might start by asking that they do before going after those comparatively unobtrusive knitters."[63] In both of these cases, textile crafts in public are weighed in relation to public social media use on mobile devices. Indeed, is it not possible that the increasing visibility (and acceptability, despite the opinions of these columnists) of sewing and knitting in public spaces in recent years may be at least partly attributable to the increased tolerance of other personal activities, including mobile media use, in such contexts?

In describing the sewing circle and space, Bratich explains that "these affinity circles traditionally existed in the margins (sometimes literally as corners and backrooms of homes)."[64] Now one could say they are in plain sight as online social circles manifest themselves in public through the craftlike physical practices of the mobile device. Knitting in public, known as "KIP" among knitters, and performing other forms of domestic labor and crafting in public can function as a form of political action and protest. World Wide Knit in Public Day is held every June in multiple countries as a way of making visible knitting's popularity as well as its social aspects. Bratich describes these practices as the "social home," which entails

the domestic sphere's practices physically coming out into public view, and the recognition that the home was always a site of convergence between social relationships and cultural economies. The social home acknowledges the oppressive conditions for women in domesticated situations, like gender domination and the exploitative reproduction of labour. … Craft-work, as part of this social home, now brings with it all the histories of affinity circles and powers activated and suppressed within the domestic sphere.[65]

Perhaps the greatest single example of this transition of crafts from the domestic to the public sphere as a form of protest and political action is the Pussyhat Project founded in November 2016. Primarily through online collaboration coordinated by the project, within weeks thousands of people knitted, crocheted, and sewed pink hats with ears to be worn by participants in the January 2017 Women's March on Washington (and elsewhere) to protest the inauguration of Donald Trump as president of the United

States.[66] Such visibility of the activities and skills of the private, domestic sphere may be at play in criticisms of mobile touchscreen culture and parallel attempts to supersede it with touchless, augmented, or virtual visual spaces. As Classen has argued, home handicraft gained popularity in the eighteenth and nineteenth century in part because it was "the site of a feminine tactics; both in its promotion of a tactually oriented aesthetic based on traditional women's work and in its manipulation of gender norms to allow women their own space."[67] Mobile and social media not only help bring craft practices out of the home and into public through the organization of knit-ins and other events, but, as was argued in chapter 3, they also do so through devices and interfaces designed to emulate aspects of craft work. This is an eventuality that extends beyond Bratich's claims. If Bratich sees the public aspect of networked communities as an opportunity to fight oppression of the home, however, then the embodied performances elicited by these devices and networks in public spaces can also appear a threat. The device brings the private, domestic, and tactile into confrontation with the public and visual in the very nature of the physical activity it produces, bringing into view actions associated with the gendered space of the knitting circle, sewing circle, and quilting bee.

## Virtual Returns

Touchscreen interfaces may be everywhere, but the dream of hands-free and invisible media use persists, at least within consumer technologies industries and much of the news reporting on them. In praising Google Glass in 2013, *Wired* columnist Matt Honan complained: "Phones separate us from our lives in all sorts of ways. Here we are together, looking at little screens, interacting (at best) with people who aren't here. Looking at our hands instead of each other."[68] While hands-free communication presents clear advantages under circumstances where the hands are needed for other activities, such as driving, it is worth examining why this perceived necessity of disengaging the hands from media continues beyond those bounds. Interaction without touch is the enduring fantasy of future digital interface. Voice recognition, immersive environments, AR, VR … in all of these, the measure of success is the extent to which one can interact with devices and the virtual environments they can produce without having to touch anything. Amid the success of mobile touchscreen media since

2007, there has been a rising discourse of AR and VR technologies as the next frontier of networked digital culture. As a teleological construction, the argument is flawed, since both (but especially VR) had already been promoted heavily as the logical next step in networked digital interface in the late twentieth-century, during the rise of the Internet. For example, in 1991 Howard Rheingold predicted that the "full impact" of VR would be felt by 2011.[69] Like the dot-com bust, however, VR lost much of its technological and social appeal in the early twenty-first century. The cognitive and perceptual mobility that VR promised became closely tied instead to the mobility of body and device through wireless networked environments as cell phones became smartphones and data networks became faster and more robust, making wireless connectivity a reliable option.

The recent return of interest in VR, sometimes tied to—or confounded with—AR, therefore takes on a different significance in a cultural context dominated by mobile media. In 2016, Facebook's Oculus Rift, Samsung's Gear VR, HTC's Vive, and Sony's PlayStation VR all appeared on the consumer market. While promotion of VR still hinges on that technology's original ideas of alternative, simulated environments that minimize, obscure, or deny the body's place in physical reality, it gains new valence in VR's role as an interface alternative to tactile mobile media. As with its earlier versions, VR is still touted as a liberating experience—even as its hardware can contain and constrict parts of the body, particularly the head and eyes—thereby marking it as a potential escape from the supposedly compromised state of mobile media as a distracting and diminishing presence in the physical world of the everyday. Such promotion of VR has often been couched in terms that return to the mind/body split and hegemonic sensorial discourses outlined previously. VR represents a return to attention, to the dominance of the visual, to the idea of a seamless, integrated world/universe for the subject, even if that world/universe is paradoxically produced by minimizing sensory perception of the physical world immediately around the user. In other words, the promise of VR and other systems relying on interactive eyewear is that they will move away from the distracted, feminizing effect of mobile media devices toward an upward, outward gaze more closely associated with masculinity and action in the field. This may be discerned, for example, in the promotion of VR for hardcore gaming typically associated with young men. More drastic, however, is the theorization of virtual reality as a technological advance that eliminates the mind/body split to afford pure consciousness, recalling Haraway's "god-trick of seeing everything from nowhere."[70] As Anne Balsamo explains, VR "is an illusion of control over reality, nature, and especially over the unruly, gender- and race-marked, essentially mortal body. … With virtual reality we are offered the vision of a body-free universe."[71]

While the haptic has always been represented in ideas of VR by data gloves, joysticks, and other prosthetic devices, VR's emphasis has customarily been placed on vision. Even in the most sensitive and critical analyses of these systems, the visual has taken precedence almost to the exclusion of other senses. Ken Hillis has traced the relationship between VR and the longer tradition of privileging sight and the mind. For him, VR is tied to ideas of producing an ideal, transparent world. As he points out, "[The idea that] our animalistic and all too finite physical bodies are thought secondary to our minds and representational forms [is] a dynamic that is built in to [sic] virtual technologies."[72] Accordingly, after explaining that subjects in immersive virtual environments "don a head-mounted display (HMD) and may wear exoskeletal devices such as a vibro-tactile feedback glove or use a joystick," Hillis discusses "optical technologies such as VR."[73] Even in the case of using VR gloves to feel or manipulate depicted objects in these environments, however, we should remember that the interface may produce an image of hands on the screen, rather than allowing our physical hands to remain visible within the VR environment.

Because we have already seen how gender, race, and other social markers function as signs that need not require a body, this escape of VR is potentially no more than a reassertion of a hegemonic vision/mind pairing, where vision and mind indeed become one. This merely perpetuates deeply embedded power structures, structures that bodily aspects of mobile tactile media can challenge in public and domestic space by making immaterial labor practices more visible. In their 1998 research on digital media as tools for remediation regularly reliant on prior media forms, Jay David Bolter and Richard Grusin demarcate a relationship between hypermedia and transparent media that may be useful here. Hypermedia, such as CD-ROMs and the World Wide Web, are "explicit acts of remediation: they import earlier media into a digital space in order to critique and refashion them." Transparent media—and Bolter and Grusin point to VR and VR games as their examples—"seek to get to the real by bravely denying the fact of mediation."[74] In the "excess" of images or conflicting cues in hypermedia, "the viewer experiences [them] not through an extended and unified gaze, but through directing her attention here and there in brief moments. The experience is one of the glance rather than the gaze," Bolter and Grusin point out. "Transparent technologies try to improve on media by erasing them," even as they remain inextricably tied to them. Virtual reality, for example, relates to the gaze of cinema, according to Bolter and Grusin.[75] Within Bolter and Grusin's theory, mobile touchscreen media could be said to occupy one pole, with most VR applications occupying the other. The touchscreen practices examined in this book relate to the conditions of hypermedia, from refashioning to "here and there" glances, in part through their ties to earlier relationships between digital and craft cultures; conversely, VR represents the recuperation of the undistracted gaze. Indeed, Bolter and Grusin assert that "new media contribute new strategies of transparency that would seem to reinforce the dissecting male gaze ... [yet] through strategies of hypermediacy, new media refashion the normative gaze and its implied views of male and female identity."[76]

Virtual reality remains more a myth than a set of technological and communicative parameters, however. In her work on the relationship between VR and the body, Diane Gromala astutely observes that the paradox of VR is the belief in achieving a sense of disembodiment by resorting to devices and processes centered on the user's body. VR "exists as a mythopoeic cultural phenomenon and as an experience through which notions of subjectivity flow and collide," she claims.[77] "Most frequently, VR is understood in terms of the experiential, inevitably referred to as the sensation of so-called disembodiment."[78] For Gromala, this myth of disembodiment, so clearly countered by the limits of VR technology, attempts nevertheless to overturn the mind/body split by simply evacuating the body from consciousness. What makes VR interesting, she claims, is how it might confound the mind/body division. While it can be said that every conscious activity already confounds the two, VR may provide an experience that illuminates this. Otherwise, it remains a technology perpetuating existing imbalances in power by making them "invisible."[79] As Cheris Kramarae wrote about VR in the late 1990s: "It seems to me that the potential major change here [with VR] is some erosion of the classic body/mind split."[80] But "in many respects, it's back to the future. It is not a revolution as much as a little shifting around in the play plot," she laments, explaining, "I can imagine the man of the house putting on his gloves and headset and visiting Costa

Rica, while the woman of the house stays in the real house, changing the diapers and fixing the meals."[81]

From its design to its marketing and reception, Glass, Google's notorious AR experiment in embedding mobile computing in eyewear, remains a critical case in the study of the social ramifications of wireless network interfaces and personal portable device practices. Glass engages the binaries of masculine and feminine, mind and body, vision and touch, and attention and distraction. It represents an alternative model of mobile media that, despite intense promotion and positive anticipation in the press and social media, was not only rejected, but openly ridiculed and reviled. Most telling, perhaps, is the device's shift of mobile interface from touch to vision, the ideological assumptions on which that shift played, and the response to that shift's implications for the social dynamics of public spaces.[82]

Google announced the Glass project in April 2012 as "technology that keeps you connected when you want to be, then gets out of your way … [and] that helps you explore and share your world, *putting you back* in the moment."[83] The accompanying promotional video, entitled "One Day," presents the Glass view of what appears to be a young white male as he spends his day navigating Manhattan while wearing the eyewear. Image and text information overlays the interior and exterior spaces he travels through, presenting route information, the location of a friend, and other details. The man interacts with the display via his voice, at one point telling Glass to take a photo in the street and share it on his social media accounts. The video ends with this unseen protagonist on a rooftop in a video phone call with a young, apparently white woman named Jessica. Jessica's small image in the frame shows her in a confining interior space. "Want to see something cool?" he asks her, then shares his skyline sunset view as he plays a ukulele (figure 4.2). "One Day" literally sets in motion a twenty-first-century urban male fantasy corresponding with the nineteenth-century Paris flâneur first described by Charles Baudelaire and picked up in the twentieth century by Walter Benjamin. As originally theorized, the flâneur is a figure who enjoys a modern vision, the city his spectacle, as he navigates its spaces. He holds a privileged position as the metropolis performs for his amusement. While the idea of the flâneur has been applied with varying success to the circulation of a female gaze within modern commercial spaces,[84] Griselda Pollock has argued that the flâneur is exclusively male, bearing a "detached observing gaze, whose possession

**Figure 4.2**
Frame from the final shot of the "One Day" Google Glass promotional video, 2012.

and power is never questioned as its basis in the hierarchy of the sexes is never acknowledged."[85] Tied to the café and consumer culture of Hausmannian Paris, the flâneur can serve as an archetypal example of the hegemony of the male gaze in modernity, even as that gaze might be under new pressures in the context of the rapidly changing social norms of the metropolis. According to Pollock, flânerie represents how "the social spaces of the city were reconstructed by the overlaying of the doctrine of separate spheres on to the division of public and private which became as a result a gendered division."[86] This condition is duplicated in the "One Day" video, where the male Glass wearer circulates freely through public spaces, while his female counterpart appears ensconced in a room and literally remains dependent on his view of the world.

Google's description of its selected Glass clients as "Explorers" and its designation of the urban workshop spaces for them as "basecamps" further emphasized a venturing masculine gaze that expanded from the gaze of the flâneur to that of the conqueror-colonizer. Google's Glass website explained that the device "frees you to look up and engage with the world around you rather than look down and be distracted from it."[87] When promoting Glass at a 2013 TED conference, Brin emphasized its psychological effect and social impact by contrasting it to handheld touchscreen devices. "In addition to potentially socially isolating yourself, when you're, you know, out and about looking [down] at your phone, uh, I feel it's kind of emasculating," he remarked. "You know, you're standing around there and you're just like rubbing this featureless piece of glass."[88] Brin's description of a loss of masculinity through touchscreen use drew some headlines, but was quickly forgotten (perhaps because the version of the talk eventually posted on the TED website excised those remarks). Nevertheless, his comments indicate the identity issues at play in media interface practices "out and about." Like Agar and Honan, Brin believes portable touchscreen devices precipitate an alienating, averted gaze, where one no longer looks others "in the eye." This suggests not only a state of distraction but, in looking down, can also suggest submission in contrast to the self-possessing dominance of the upright, level gaze. Additionally, smartphones place an emphasis on direct touch over remote command (speech activation, in the case of Glass) in the device interface, thereby privileging the baseness of bodily contact over distanced control through view and speech. As Brin rightly infers, though does not explicitly state, all of these traits are ascribed historically to femininity. By replacing these with spoken commands, upright posture, and a steady, direct gaze outward, Glass becomes the means of reestablishing masculinity in the public media sphere. "When we developed Glass, we thought really about 'Can we make something that frees your hands?,'" he explained to his TED audience, "and we want something that frees your eyes."[89]

By embedding the screen in eyewear and privileging voice prompts (though also allowing touch prompts by tapping the frame edge), Glass recovered the digital interface from the feminizing connotations of embodied handiwork, striving toward the self-flattery described by Merleau-Ponty. The shift is implied in Isabel Pedersen's observation that "wearable media sits midway between media you carry … and media you become." She cites developments in contact lens displays that are similar to those of Glass: "Instead of *looking at* handheld devices, we will *wear* see-through displays."[90] Perhaps it should be no surprise, then, that one of the most persistent criticisms of Glass was that it facilitated a voyeuristic and intrusive gaze. If Glass was meant to be transparent, allowing the wearer to interact visually with the surrounding environment as one would while wearing ordinary eyeglasses, it was instead a conspicuous object in ways the handheld mobile device is not.[91] It placed the device directly within the view of the other, intervening between the wearer and any companion or counterpart. Glass was perceived as a threat to privacy and an aid to voyeurism as wearer-users might retrieve information and record images of the visual field without the knowledge of those around them. That this perception downplays the real possibility of similar actions with mobile touchscreen devices or hidden cameras points to the sometimes significant gap between a technology's capabilities and applications, and how it is perceived. Critics referred to Explorers as "Glassholes," and Google offered Glass users etiquette tips for public situations that included "Don't ... be creepy or rude" by doing things such as "standing alone in the corner of a room staring at people while recording them."[92]

Although handheld touchscreen devices have been an unmitigated success for mobile media, and Glass an utter failure, the fight for the outward, attentive gaze continues in the promotion of VR headgear. If the portable device cannot be beaten, it can be absorbed. Google thus followed Glass with Cardboard, the VR system that adopts a lowly do-it-yourself aesthetic. Cardboard wraps a standard smartphone in a cardboard frame (figure 4.3) that, when combined with specially designed videos playing on the screen, gives the user a sense of immersion in a simulated 3-D environment. Google claims Cardboard is immersive VR "for everyone," implying that other systems are not.[93] Competing electronics and software companies, such as Samsung, have followed with variations of VR that also rely on, or incorporate, portable touchscreen devices.[94]

Recent versions of VR—especially those that rely on mobile devices for their screens—may not adhere as closely to its earlier driving myth of disembodiment, but they can challenge mobile devices by incorporating them fully into the VR system. "A whole virtual-reality industry has grown around a device you already have in your pocket," one tech writer has explained. "Right now, your phone is the single most important device in VR."[95] The VR frames that surround the mobile device—essentially blinders of varying technological sophistication—become prostheses that make it—and its standard uses—"disappear." It no longer occupies the hand in the same way, no longer speaks to that intimate fit, and no longer serves as the frame for mixing, matching, and assembling material. In other words, both user and device become locked into a relationship that allows the primacy of the visual to reemerge in images that are literally untouched.

Theorization of computing in the late twentieth century elaborates on gender-based differences in approaching digital technology that may still

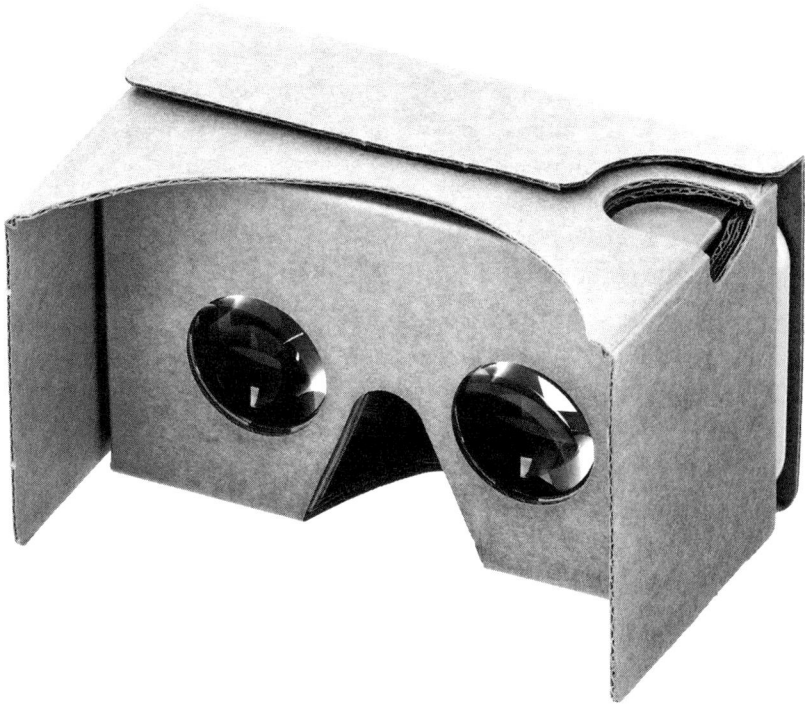

**Figure 4.3**
Google Cardboard headset, with the edge of an enclosed smartphone visible at right, 2015. Photo by Evan Amos.

resonate in these discrepancies between the form and function of handheld touchscreen devices and VR systems. Researching children in the 1980s, Sherry Turkle developed the idea of "hard" versus "soft" mastery in computing approaches. Hard masters attempt "the imposition of will over the machine through the implementation of the plan," according to Turkle, while soft masters are more interactive. "Like the *bricoleur*, the soft master works with a set of concrete elements ... arranging and rearranging these elements, working through new combinations."[96] Turkle concluded that boys were generally hard masters in their interactions with computers, while girls tended to be soft masters. More recently, Judy Wajcman has revisited Turkle's categories in exploring how such gendered approaches may be culturally instilled. Wajcman states that "computer expertise is defined as hard mastery [while] soft mastery is culturally constructed as inferior."[97]

VR engages the notion of hard mastery, fostering the idea that one can set out on a straight path. It emulates the linearity embedded in Renaissance perspective and that system's ideal viewing point and corresponding vanishing point of infinitude.[98] The same linearity emerges in Google's "One Day" Glass video with its concluding shot of a man gazing at the horizon and its vanishing point of the setting sun. It is a sense of mastery over material without getting one's hands dirty. The portable touchscreen is all about materials in hand, however. It is not about linearity, but about sampling, about putting things together through multiple apps, platforms, and windows. It is a pieced-together exploration mastered by the fingers as much as the eye. Indeed, to an outside observing eye, it might appear to be a hodge-podge or some form of chaos.

If the ultimate desire of VR is to create a context where the screen and interface seem to disappear, touchscreen mobile devices are about their conspicuous presence. Not only are these objects visible in public spaces, constantly popping up here and there as users take them out and put them away, but their presence is further indicated by the rapid movements of fingers across their surfaces. They do not suggest a seamlessness to space or experience, but rather a negotiated assemblage of experience, along the lines of the soft mastery described by Turkle. That mobile media expertise may be associated with practices and techniques that historically have been marked as feminine, domestic, and unskilled introduces the possibility for a major shift in our approach to digital media, as well as an opportunity for significant social change in our relationships to technology and identity more generally. If and how these are realized remain to be seen. The false promises of virtual reality and similar systems as seamless alternatives, however, may remind us of continued efforts to undermine that potential with interfaces that would obscure our body's relationship to both technology and the ideological assumptions that course through it.

# Notes

## Introduction

1. See, for example, numerous essays collected in Gail Dines and Jean M. Humez, eds., *Gender, Race, and Class in Media: A Critical Reader*, 3rd ed. (Thousand Oaks, CA: SAGE, 2011); Mary Celeste Kearney, ed., *The Gender and Media Reader* (New York: Routledge, 2011); and Niall Richardson and Sadie Wearing, *Gender in the Media* (Basingstoke: Palgrave Macmillan, 2014). Since most TV series, films, games, and even greeting cards are distributed through digital networks accessible to mobile media, nearly any treatment of gender in contemporary media content implicitly concerns mobile media content.

2. Jay Bolter and Diane Gromala, *Windows and Mirrors: Interaction Design, Digital Art, and the Myth of Transparency* (Cambridge, MA: MIT Press, 2003), 27.

3. Esquire Editors, "Are You Man Enough for a Pink iPhone 6s?," *Esquire*, September 26, 2015, http://www.esquire.com/lifestyle/cars/news/a38262/new-iphone-6s-colors/ (accessed January 7, 2017).

4. Lisa Gitelman, *Always Already New: Media, History, and the Data of Culture* (Cambridge, MA: MIT Press, 2006), 7.

5. Another, related path of exploration might be networked mobile media's ties to telephony and the gendered labor and communicative practices that have marked that medium's history.

6. See Andrew Butterfield, Gerard Ekembe Ngondi, and Anne Kerr, eds., *A Dictionary of Computer Science*, 7th ed. (Oxford: Oxford University Press, 2016). Additional examples can be found in Jack Bratich, "The Digital Touch: Craft-Work as Immaterial Labour and Ontological Accumulation," *Ephemera* 10, no. 3–4 (2010): 303.

7. For example, while this book draws an analogy between the mobile screen and the handloom, Francesco Casetti suggests the metaphor of the scrapbook is apt, since networked users "construct images of themselves in the first person, by

assembling photos, texts and comments often lifted from elsewhere." The frequent use of the term "cutting and pasting" to describe such activities would support his choice. See Francesco Casetti, "What Is a Screen Nowadays?," in *Public Space, Media Space*, ed. Chris Berry, Janet Harbord, and Rachel O. Moore (Basingstoke: Palgrave Macmillan, 2013), 26.

8. Lisa Gitelman's work is an important example here, as she has sought to dissolve the false, and sometimes arbitrary, divides that rise between old and new media. See Lisa Gitelman and Geoffrey B. Pingree, eds., *New Media, 1740–1915* (Cambridge, MA: MIT Press, 2003); and Lisa Gitelman, *Always Already New: Media, History, and the Data of Culture* (Cambridge, MA: MIT Press, 2006).

9. Stella Minahan and Julie Wolfram Cox, "Stitch'nBitch: Cyberfeminism, a Third Place and the New Materiality," *Journal of Material Culture* 12, no. 1 (2007): 6. Emphasis in original.

10. Alissa Haight Carlton and Kristen Lejnieks, *Block Party: The Modern Quilting Bee* (Lafayette, CA: C&T Publishing, 2011), 9.

11. Bratich, "The Digital Touch," 307.

12. The Craftivist Collective (https://craftivist-collective.com/) is an example.

13. Kirsty Robertson explores the paradox of craftivists using online media in "Rebellious Doilies and Subversive Stitches: Writing a Craftivist History," in *Extra/Ordinary: Craft and Contemporary Art*, ed. Maria Elena Buszek (Durham, NC: Duke University Press, 2011), 184–203.

14. James Bridle, "The New Aesthetic and Its Politics," June 12, 2013, http://booktwo.org/notebook/new-aesthetic-politics/ (accessed January 7, 2017).

15. http://www.microrevolt.org/knitPro/.

16. http://www.spritestitch.com.

17. Richard Trenholm, "Wearables at War: How Smart Textiles Are Lightening the Load for Soldiers," CNet, March 11, 2015, https://www.cnet.com/news/wearables-at-war-how-smart-textiles-are-lightening-the-load-for-soldiers/ (accessed January 7, 2017).

18. For the technical side of these fabrics, see Gilsoo Cho, ed, *Smart Clothing: Technology and Applications* (Boca Raton: CRC Press, 2010); the social uses and implications of smart clothing are considered in Susan Elizabeth Ryan, *Garments of Paradise: Wearable Discourse in the Digital Age* (Cambridge, MA: MIT Press, 2014).

19. Johanna Drucker, *Graphesis: Visual Forms of Knowledge Production* (Cambridge, MA: Harvard University Press, 2014), 8.

20. Branden Hookway, *Interface* (Cambridge, MA: MIT Press, 2014), 16.

21. Wendy Hui Kyong Chun, *Programmed Visions: Software and Memory* (Cambridge, MA: MIT Press, 2011), 66–67. Here Chun is drawing on descriptions of ideology by Friedrich Engels and Louis Althusser.

22. Drucker, *Graphesis*, 146.

23. Christian Ulrik Andersen and Søren Bro Pold, "Interface Criticism: Aesthetics Beyond Buttons," in *Interface Criticism: Aesthetics Beyond Buttons*, ed. Christian Ulrik Andersen and Søren Bro Pold (Aarhus: Aarhus University Press, 2011), 9.

24. Alexander R. Galloway, *The Interface Effect* (Malden, MA: Polity, 2012); and Hookway, *Interface*.

25. Galloway, *The Interface Effect*, vii–viii; Hookway, *Interface*, 6–7.

26. Galloway, *The Interface Effect*, vii.

27. Hookway, *Interface*, ix.

28. Ibid., 39–40.

29. Ibid., ix.

30. Andersen and Pold, "Interface Criticism," 7. See also Dragana Antic and Matthew Fuller, "The Computation of Space," in *Interface Criticism*, ed. Andersen and Pold: 130–142.

31. This points to not only the biopolitics of Michel Foucault, but also Henri Lefebvre's theory of space as social production in and of itself. Both of these theories have contributed to the perspective and approach of this book. See Michel Foucault, *Discipline and Punish: The Birth of the Prison*, trans. Alan Sheridan (New York: Vintage, 1995); and Henri Lefebvre, *The Production of Space*, trans. Donald Nicholson-Smith (Oxford: Blackwell, 1991).

32. Malcolm McCullough, *Ambient Commons: Attention in the Age of Embodied Information* (Cambridge, MA: MIT Press, 2015).

33. Judith Butler, *Bodies That Matter: On the Discursive Limits of Sex* (New York: Routledge, 1993), 9.

34. Judith Butler, *Gender Trouble: Feminism and the Subversion of Identity* (New York: Routledge, 1990), 136.

35. Elizabeth Grosz, *Volatile Bodies: Toward a Corporeal Feminism* (Bloomington: Indiana University Press, 1994), 14. See also Simone de Beauvoir, *The Second Sex*, trans. Constance Borde and Sheila Malovany-Chevallier (London: Jonathan Cape, [1949] 2009); and Iris Marion Young, "A Phenomenology of Feminine Body Comportment Motility and Spatiality," *Human Studies* 3, no. 2 (1980): 137–156.

36. David Gelernter, *Machine Beauty: Elegance and the Heart of Technology* (New York: Basic Books, 1998), 89. Emphasis in original.

37. Wendy Hui Kyong Chun, *Control and Freedom: Power and Paranoia in the Age of Fiber Optics* (Cambridge, MA: MIT Press, 2006); Wendy Hui Kyong Chun, *Updating to Remain the Same: Habitual New Media* (Cambridge, MA: MIT Press, 2016); Matthew Fuller and Andrew Goffey, *Evil Media* (Cambridge, MA: MIT Press, 2012); Alexander R. Galloway, *Protocol: How Control Exists after Decentralization* (Cambridge, MA: MIT Press, 2004); Lisa Gitelman, *Paper Knowledge: Toward a Media History of Document* (Durham, NC: Duke University Press, 2014); Hookway, *Interface*; Tung-Hui Hu, *A Prehistory of the Cloud* (Cambridge, MA: MIT Press, 2015); Adrian Mackenzie, *Cutting Code: Software and Sociality* (New York: Peter Lang, 2006); and Jonathan Sterne, *MP3: The Meaning of a Format* (Durham, NC: Duke University Press, 2012).

38. Examples include Eric Gordon and Adriana de Souza e Silva, eds., *Net Locality: Why Location Matters in a Networked World* (Malden, MA: Wiley-Blackwell, 2011); Rob Kitchin and Martin Dodge, eds., *Code/Space: Software and Everyday Life* (Cambridge, MA: MIT Press, 2011); Jason Farman, *Mobile Interface Theory: Embodied Space and Locative Media* (New York: Routledge, 2012); Nanna Verhoeff, *Mobile Screens: The Visual Regime of Navigation* (Amsterdam: Amsterdam University Press, 2012); and Marianne Van Den Boomen, *How Metaphors Matter in New Media: Transcoding the Digital* (Amsterdam: Amsterdam University Press, 2015).

39. Frank Bentley and Edward Barrett, *Building Mobile Experiences* (Cambridge, MA: MIT Press, 2012), 7.

40. See Joe Marshall and Paul Tennent, "Mobile Interaction Does Not Exist," in *CHI2013 Changing Perspectives: Extended Abstracts: The 31st Annual CHI Conference on Human Factors in Computing Systems*, ed. Patrick Baudisch, Michel Beaudouin-Lafon, and Wendy E. Mackay (New York: Association for Computing Machinery, 2013), 2069–2078.

41. Among these are Lev Manovich, *The Language of New Media* (Cambridge, MA: MIT Press, 2001); Lev Manovich, *Software Takes Command* (London: Bloomsbury, 2013); Noah Wardrip-Fruin and Nick Monfort, eds., *The New Media Reader* (Cambridge, MA: MIT Press, 2003); and Eduardo Navas, *Remix Theory: The Aesthetics of Sampling* (New York: Springer, 2012).

42. Lev Manovich, "New Media from Borges to HTML," in *The New Media Reader*, ed. Noah Wardrip-Fruin and Nick Montfort (Cambridge, MA: MIT Press, 2001), 16.

43. Ibid., 22–23.

44. Eduardo Navas, "Culture and Remix: A Theory on Cultural Sublation," in *The Routledge Companion to Remix Studies*, ed. Eduardo Navas, Owen Gallagher, and xtine burrough (New York: Routledge, 2014), 116.

45. Bruce Metcalf, "Replacing the Myth of Modernism," in *Neo-Craft: Modernity and the Crafts*, ed. Sandra Alfody (Halifax: Press of the Nova Scotia College of Art and Design, 2007), 4–32.

46. Eric McLuhan, "Arachne or Penelope: Queen of the Net, Mistress of the Web?," chap. 1 in *Electric Language: Understanding the Present* (Toronto: Stoddart, 1998), 7.

47. Malcolm McCullough, *Abstracting Craft: The Practiced Digital Hand* (Cambridge, MA: MIT Press, 1996), 21–22.

48. Another example of such an understanding of craft would be the call to "flexible specialization" through "craft forms of production" in Michael J. Piore and Charles F. Sabel, *The Second Industrial Divide: Possibilities for Prosperity* (New York: Basic Books, 1984), 17.

49. Brenda Danet, *Cyberpl@y: Communicating Online* (Oxford: Berg, 2001), 243–244 and 257–258.

50. Ibid., 244.

51. Sadie Plant, *Zeroes + Ones: Digital Women and the New Technoculture* (New York: Doubleday, 1997); Bratich, "The Digital Touch"; Jack Bratich and Heidi Brush, "Fabricating Activism: Craft-Work, Popular Culture, Gender," *Utopian Studies* 22, no. 2 (2011): 233–260; and Minahan and Cox, "Stitch'nBitch."

52. L. F. Menabrea, "Sketch of the Analytical Engine Invented by Charles Babbage, Esq.," trans. Ada Lovelace, in *Scientific Memoirs, Selected from the Transactions of Foreign Academies of Science and Learned Societies, and from Foreign Journals*, vol. 3, ed. Richard Taylor (London: Richard and John E. Taylor, 1843), 696. Emphasis in original removed.

## 1 Woven Memory

1. There are scores of examples. To provide a few: M. V. Wilkes, *Automatic Digital Computers* (London: Methuen, 1956), 6–9; Albert Paul Malvino and Donald P. Leach, *Digital Principles and Applications*, 2nd ed. (New York: McGraw-Hill, 1975), 1–3; René Moreau, *The Computer Comes of Age: The People, the Hardware, the Software*, trans. J. Howlett (Cambridge, MA: MIT Press, [1981] 1984), 13–14; Martin Campbell-Kelly and William Aspray, *Computer: A History of the Information Machine* (New York: Basic Books, 1996), 55–57; James Essinger, *Jacquard's Web: How a Hand-Loom Led to the Birth of the Information Age* (Oxford: Oxford University Press, 2004); Neil Barrett, *The Binary Revolution: The Development of the Computer* (London: Weidenfeld & Nicolson, 2006), 20–22; Ian Watson, *The Universal Machine: From the Dawn of Computing to Digital Consciousness* (New York: Copernicus Books, 2012), 27–28; and Joel Shurkin, *Engines of the Mind: A History of the Computer* (New York: W. W. Norton, 1984), 58–59.

2. Jacquard's invention was an improvement on less complex versions of punched-card weaving systems devised in France by Basile Bouchon in 1725 and Jacques de Vaucanson in 1745. See Eric Broudy, *The Book of Looms: A History of the Handloom*

*from Ancient Times to the Present* (Hanover, NH: University Press of New England, 1979), 134.

3. In France's brocade industry in the eighteenth century, young women sometimes performed this physically demanding job. Discussions of allowing drawgirls into weavers' guilds, however, brought resistance from master weavers of the time. See Daryl M. Hafter, "The 'Programmed' Brocade Loom and the Decline of the Draw-Girl," in *Dynamos and Virgins Revisited: Women and Technological Change in History*, ed. Martha Moore Trescott (Metuchen, NJ: Scarecrow Press, 1979), 49–66; and Daryl M. Hafter, *Women at Work in Preindustrial France* (State College: Pennsylvania State University Press, 2010), 197–198.

4. Julie Holyoke, *Digital Jacquard Design* (London: Bloomsbury, 2013).

5. George Caffentzis is one of the few scholars to make the connection. He explains: "[An] important reason for the neglect of Babbage's Engines was that neither Babbage, nor Marx, nor anyone else at the time saw the essential connection between computation and *all* forms of the labor process, even though the key was staring Babbage and Marx in the face all along. That key was the Jacquard loom." George Caffentzis, *In Letters of Blood and Fire: Work, Machines, and the Crisis of Capitalism* (Chicago: PM Press, 2013), 195.

6. This summary of the Analytical Engine's construction is based on the detailed description of its physical attributes and functioning provided in Allan G. Bromley, "Charles Babbage's Analytical Engine, 1838," *Annals of the History of Computing* 4, no. 3 (July 1982): 196–217.

7. Charles Babbage, *Passages from the Life of a Philosopher* (London: Longman, Green, Longman, Roberts, & Green, 1864), 116–117.

8. John von Neumann, "First Draft of a Report on the EDVAC," in *Great Papers in Computer Science*, ed. Phillip Laplante (St. Paul, MN: West Publishing, 1996), 211. Emphasis in original.

9. Wendy Hui Kyong Chun, "The Enduring Ephemeral, Or the Future Is a Memory," *Critical Inquiry* 35 (Autumn 2008): 154, reprinted in "The Enduring Ephemeral, Or the Future Is a Memory," in *Media Archaeology*, ed. Erkki Huhtamo and Jussi Parikka, 188. See also Gerd Gigerenzer and Daniel G. Goldstein, "Mind as Computer: Birth of a Metaphor," *Creativity Research Journal* 9, no. 2–3 (1996): 131–144.

10. Hannah B. Higgins, *The Grid Book* (Cambridge, MA: MIT Press, 2009), 240.

11. Ada Lovelace has been a leading figure in histories and theories of computing developing out of 1990s cyberfeminism. Sadie Plant made Lovelace's life and work the organizing trope in *Zeroes + Ones: Digital Women and the New Technoculture* (New York: Doubleday, 1997), her theorization of the relationship between gender and technology. Alex Galloway has called this the "recuperationist school" of cyberfeminism, which highlights the achievements of Lovelace and subsequent women in

computing, such as the twentieth-century American computer developer Grace Hopper. See Alex Galloway, "A Report on Cyberfeminism: Sadie Plant Relative to VNS Matrix," *Switch* 4, no. 1, http://switch.sjsu.edu/web/v4n1/alex.html (accessed January 7, 2017). In this chapter Lovelace is most significant for her cultural interpretation of the Analytical Engine, all the more important given her work on its programming. Because this study concerns gendered popular culture and labor practices as they have impacted the design and function of contemporary digital devices and interfaces, however, the individual roles played by many women in computing history have largely been set aside. For those pivotal contributions, see Janet Abbate, *Recoding Gender: Women's Changing Participation in Computing* (Cambridge, MA: MIT Press, 2012); and Autumn Stanley, *Mothers and Daughters of Invention: Notes for a Revised History of Technology* (Metuchen, NJ: Scarecrow Press, 1993), 642–745.

12. L. F. Menabrea, "Sketch of the Analytical Engine Invented by Charles Babbage, Esq.," trans. Ada Lovelace, in *Scientific Memoirs, Selected from the Transactions of Foreign Academies of Science and Learned Societies, and from Foreign Journals*, vol. 3, ed. Richard Taylor (London: Richard and John E. Taylor, 1843), 696. Emphasis in original.

13. Ibid., 698.

14. Kax Wilson, *A History of Textiles* (Boulder, CO: Westview Press, 1979), 263.

15. See Anne Isabella Byron's journal entry of July 17, 1834—in which she illustrates the punched-card system of a Coventry mill and describes the child labor conditions there—as reproduced in Essinger, *Jacquard's Web*, 138–139.

16. Menabrea, "Sketch of the Analytical Engine Invented by Charles Babbage, Esq.," 706. Emphasis in original.

17. Natalie Rothstein, "The Introduction of the Jacquard Loom to Britain," in *Studies on Textile History: In Memory of Harold B. Burnham*, ed. Veronika Gervers (Toronto: Royal Ontario Museum, 1977), 282.

18. Joy Spanabel Emery, *A History of the Paper Pattern Industry: The Home Dressmaking Fashion Revolution* (London: Bloomsbury Academic, 2014), 19–28.

19. Tim Putnam, "The Sewing Machine Comes Home," in *The Culture of Sewing: Gender, Consumption, and Home Dressmaking*, ed. Barbara Burman (Oxford: Berg, 1999), 269. See also Ruth Brandon, *A Capitalist Romance: Singer and the Sewing Machine* (Philadelphia: Lippincott, 1977).

20. For this history, see Kevin L. Seligman, *Cutting for All!: The Sartorial Arts, Related Crafts, and the Commercial Paper Pattern* (Carbondale: Southern Illinois University Press, 1996); and Emery, *A History of the Paper Pattern Industry*.

21. Essinger's book-length examination of Jacquard, his apparatus, and their impact on computing, for example, makes no mention of the invention's more immediate social or economic ramifications. Essinger, *Jacquard's Web*.

22. Wilson, *A History of Textiles*, 63.

23. Jennifer Harris, "A Survey of Textile Techniques," in *5,000 Years of Textiles*, ed. Jennifer Harris (London: British Museum, 1993), 19; and William Felkin, *History of the Machine-Wrought Hosiery and Lace Manufacturers* (Newton Abbot: David & Charles, [1867] 1967), 324–343. Natalie Rothstein notes that the Jacquard mechanism did not enter into common use in the Lyon textile industry until the 1810s. She attributes the delay to the popular preference for simple patterns and prints in early nineteenth-century Europe. Rothstein, "The Introduction of the Jacquard Loom to Britain," 284.

24. Other circumstances contributing to this social movement included the introduction of shearing frames with gig mills to automatically finish cloth and the "cut-up" technique of manufacturing knitwear. Woolen mills employing this technique would knit swathes of fabric that would then be cut and stitched in the same way as woven cloth, rather than knitting the articles directly. Cut-up assembly produced cheap, poorly finished articles that deteriorated quickly. A. L. Morton and George Tate, *The British Labour Movement* (London: Lawrence & Wishart, 1956), 36–38; and J. L. Hammond and Barbara Hammond, *The Skilled Labourer, 1760–1832* (New York: Harper & Row, [1919] 1970), 226–227 and 257–258.

25. *Annual Register* (London: Otridge and Son, 1812), 385, as quoted in Hammond and Hammond, *The Skilled Labourer*, 257. Hammond and Hammond discredit workplace gender shifts as a factor in Luddism, although homeworking women were involved in the movement. Nevertheless, Parliament's acknowledgment of gender and manufacturing suggests how remarkable such changes were at that time. For his part, Lord Byron argued in sympathy of workers involved in the protests. See Malcolm Kelsall, *Byron's Politics* (Sussex: Harvester Press, 1987), 38–51.

26. Charles Babbage, *On the Economy of Machinery and Manufactures* (London: John Murray, [1832] 1846), 336–337.

27. Ibid., 341.

28. Ibid., 339.

29. For an analysis of the ways automation historically has benefited the male workforce and social sphere while increasing work—particularly underpaid or unpaid work—and diminishing social opportunities for women, see Susan Strasser, *Never Done: A History of American Housework* (New York: Pantheon, 1982).

30. Judy Lown, *Women and Industrialization: Gender at Work in Nineteenth-Century England* (Minneapolis: University of Minnesota Press, 1990), 43–45.

31. For an analysis of the ideological relationship between skill and gender in labor practices, see Shirley Dex, *The Sexual Division of Labour* (Brighton: Wheatsheaf Books, 1985), 99–104 and 187–193.

32. Mary Frank Fox and Sharlene Hesse-Biber, *Women at Work* (Palo Alto, CA: Mayfield Publishing, 1984), 86. For information on collective actions, see Thomas Dublin, *Women at Work: The Transformation of Work and Community in Lowell, Massachusetts, 1826–1860* (New York: Columbia University Press, 1979), chapters 6 and 7.

33. Joel Shurkin, *Engines of the Mind: The Evolution of the Computer from Mainframes to Microprocessors* (New York: W. W. Norton, [1984] 1996), 126.

34. A particularly late example of human computers is the West Area Computing Unit of the National Advisory Committee for Aeronautics and the National Aeronautics and Space Administration. It employed many African-American women mathematicians from the 1940s into the 1960s. See Margot Lee Shetterly, *Hidden Figures: The Untold True Story of Four African-American Women Who Helped Launch Our Nation into Space* (New York: Harper, 2016).

35. David Rutland, *Why Computers Are Computers: The SWAC and the PC* (Philomath, OR: Wren, 1995), 29–30.

36. Jennifer Light, "When Computers Were Women," *Technology and Culture* 40, no. 3 (July 1999): 455–483.

37. Ibid., 471. See also W. Barkley Fritz, "The Women of ENIAC," *IEEE Annals of the History of Computing*, 18, no. 3 (1996): 13–28.

38. The multiple strands of this development emerge in Paul E. Ceruzzi, *A History of Modern Computing*, 2nd ed. (Cambridge, MA: MIT Press, [1998] 2003); Raúl Rojas and Ulf Hashagen, eds., *The First Computer: History and Architectures* (Cambridge, MA: MIT Press, 2002); and Michael Sean Mahoney, *Histories of Computing*, ed. Thomas Haigh (Cambridge, MA: Harvard University Press, 2011).

39. Paul N. Edwards, *The Closed World: Computers and the Politics of Discourse in Cold War America* (Cambridge, MA: MIT Press, 1996), 75.

40. An extensive description of the development and functioning of core memory can be found in Emerson W. Pugh, *Memories That Shaped an Industry: Decisions Leading to IBM System/360* (Cambridge, MA: MIT Press, 1984), 62–128.

41. For a detailed explanation of the configuration and operation of core memory that nevertheless remains accessible to nonspecialists, see Wilkes, *Automatic Digital Computers*, 194–209. Forrester's description can be found in Jay W. Forrester, "Digital Information Storage in Three Dimensions Using Magnetic Cores," *Journal of Applied Physics* 22, no. 1 (January 1951): 44–48.

42. C. Robert Weiser, "The Cape Cod System," *Annals of the History of Computing* 5, no. 4 (October 1983): 362–369. Those working on core memory included independent inventor Frederick Viehe, Jan Rajchman at RCA Laboratories, Munro K. Haynes at IBM, and An Wang at the Harvard Computation Laboratory. Wang's account is

available in An Wang and Eugene Linden, *Lessons: An Autobiography* (Reading, MA: Addison-Wesley, 1986), 45–61. See also Michael R. Williams, *A History of Computing Technology*, 2nd ed. (Los Alamitos, CA: IEEE Computer Society Press, 1997), 319–320; and Emerson W. Pugh, *Building IBM: Shaping an Industry and Its Technology* (Cambridge, MA: MIT Press, 1995), 209.

43. Martin Campbell-Kelly, William Aspray, Nathan Ensmenger, and Jeffrey R. Yost, *Computer: A History of the Information Machine*, 3rd ed. (Boulder, CO: Westview Press, 2014), 150.

44. William N. Papian, "A Coincident-Current Magnetic Memory Unit" (MA thesis, Massachusetts Institute of Technology, 1950).

45. Bernard Widrowitz, "The 16 by 16 Metallic Memory Array, Model 1," Report R-216, Digital Computer Laboratory, Massachusetts Institute of Technology (September 25, 1952). The use of a single example of a $16 \times 16$ array is confirmed by Maurice Wilkes, who visited the lab in August 1952. Maurice V. Wilkes, *Memoirs of a Computer Pioneer* (Cambridge, MA: MIT Press, 1985), 181.

46. Pugh, *Memories That Shaped an Industry*, 98–99. Papian envisioned the $64 \times 64$ array in his master's thesis, and examples were assembled by August 1953. In 1954, the lab produced $128 \times 128$ arrays and planned to attempt a $256 \times 256$ array, which would require hand-threading and orienting 65,536 cores. See Papian, "A Coincident-Current Magnetic Memory Unit"; and Pugh, *Memories That Shaped an Industry*, 127.

47. The laboratory's Memory Section hired another woman technician, Betty Kollet, a few months later to help with testing cores. "Bi-Weekly Report," Memorandum M-1779, Digital Computer Laboratory, Massachusetts Institute of Technology (January 2, 1953), 14 and 32; "Bi-Weekly Report," Memorandum M-1825, Digital Computer Laboratory, Massachusetts Institute of Technology (February 9, 1953), 9; and "Bi-Weekly Report," Memorandum M-2356, Digital Computer Laboratory, Massachusetts Institute of Technology (August 15, 1953), 14.

48. N. H. Taylor, "The Construction of Memory Planes for the MTC Memory," Memorandum M-2225, Digital Computer Laboratory, Massachusetts Institute of Technology (June 10, 1953), 1.

49. Wilkes offers his firsthand impressions of the development of core memory from his visits to MIT between 1950 and 1953 in Wilkes, *Memoirs of a Computer Pioneer*, 181–182.

50. As quoted in the video *Quick Facts: The World's First RAM Chip, with Inventor Bernard Widrow*, dir. Michael Whelan (Schenectady, NY: Edison Tech Center, 2009). Little more than Widrow's reminiscence and Carpenter's 2013 online obituary remain in the public record to document her contribution. See "Hilda Carpenter," *Florida Times-Union*, August 23, 2013, http://www.legacy.com/obituaries/timesunion/obituary.aspx?pid=166570014 (accessed January 7, 2017).

51. Unfortunately, there is no mention of Carpenter in Clarence G. Williams, ed., *Technology and the Dream: Reflections on the Black Experience at MIT, 1941–1999* (Cambridge, MA: MIT Press, 2001), though her obituary states that she was a long-time member of the NAACP. See "Hilda Gwendolyn Carpenter," A. B. Coleman Mortuary, http://www.abcoleman.net/sitemaker/memsol.cgi?user_id=1067889 (accessed August 14, 2016). Carpenter continued at Lincoln Laboratory after Whirlwind. A photograph in the collection of the Computer History Museum (catalog number 102622651) documents her subsequent participation on the development team of the LINC minicomputer. See http://www.computerhistory.org/collections/catalog/102622651 (accessed January 7, 2017).

52. Pugh recounts that when IBM began producing core memory planes "Occasionally the women assembling the core frames broke down and cried when a final inspection revealed a fault in the wiring that required the wires be removed and restrung. The emotion of seeing so much labor lost was too much for the conscientious people needed for this exacting work." One might also consider that, in view of the tight delivery schedules, repeated faults in wiring would likely cost an assembler her job. Pugh, *Memories That Shaped an Industry*, 111–112.

53. Fellowcrafters Guild, *Easiweaving: A Pattern Book* (Boston: Fellowcrafters, 1936), i.

54. H. Atwood Reynolds, *Complete Book of Modern Crafts* (New York: Blue Ribbon Books, 1940), 223.

55. Despite core memory's close aesthetic and material relationship to these processes, the final product would be stacked and shut away in cases in the computer room, while handloom creations were meant to be displayed, shared, and exchanged.

56. E. A. Guditz and L. B. Smith, "Vacuum and Vibration Speed Assembly of Core Memory," *Electronics* 29, no. 2 (February 1956): 214.

57. Pugh, *Building IBM*, 177 and 207–219.

58. Michael Williams, *A History of Computing Technology* (Los Alamitos: IEEE Computer Society Press, 1997), 320.

59. Stan Augarten, *Bit by Bit: An Illustrated History of Computers* (New York: Ticknor & Fields, 1984), 257.

60. Jury E. Seleznev, Jury A. Burkin, Sergei V. Kuzmin, "Ferrite Core Memory," U.S. 4161037 A, filed January 17, 1977, and issued July 10, 1979.

61. David R. Boles, John S. Davis, and Paul E. Wells, "Apparatus for Woven Screen Memory Devices," U.S. 3377581 A, filed November 12, 1963, and issued April 9, 1968; J. S. Davis and P. E. Wells, "Investigation of a Woven Screen Mass Memory System," *AFIPS* (1963): 311–326; and H. Maeda, M. Takashima, and A. J. Kolk Jr., "A

High-Speed, Woven Read-Only Memory," *AFIPS* (1965): 789–799. To avoid hand assembly, Western Electric developed "ferrite sheet" memory. A layered, meshlike product that avoided cores and bias stitching, ferrite sheet was fully adaptable to mechanized batch production. It apparently did not match the desirable combination of versatility, capacity, and speed found in core memory planes, however, and found application primarily in automated telephone switching stations, where temporary memory needs were simple, but critical. See Robert J. Chapius and Amos E. Joel, *100 Years of Telephone Switching: Electronics, Computers, and Telephone Switching, 1960–1985* (Amsterdam: IOS Press, 2003), 98.

62. Lucien V. Auletta, Herbert J. Hallstead, and Denis J. Sullivan, "Ferrite Core Planes and Arrays: IBM's Manufacturing Evolution," *IEEE Transactions on Magnetics* 5, no. 4 (December 1969): 764–774.

63. Pugh, *Memories That Shaped an Industry*, 137.

64. This was called the "Mecca wiring pattern," after the name of the task force that created it. Another, less effective array pattern reduced the array to two wires running through each core. Emerson W. Pugh, Lyle R. Johnson, and John H. Palmer, *IBM's 360s and Early 370 Systems* (Cambridge, MA: MIT Press, 1991), 186–187 and 198; and Pugh, *Memories That Shaped an Industry*, 231–233.

65. Richard Thomas DeLamarter, *Big Blue: IBM's Use and Abuse of Power* (New York: Dodd, Mead, and Company, 1986), 207.

66. Pugh, Johnson, and Palmer, *IBM's 360s and Early 370 Systems*, 208.

67. David A. Grier, *Too Soon To Tell: Essays for the End of the Computer Revolution* (Hoboken, NJ: Wiley, 2009), 91.

68. Pugh, Johnson, and Palmer, *IBM's 360s and Early 370 Systems*, 209.

69. Systems were tested for vibration, shock, acceleration, temperature, vacuum, humidity, salt fog, and electronic noise. The durability results for MIT's original configurations were disastrous, requiring a thorough redesign through which rope memory emerged as an expensive but viable option. See James E. Tomayko, "Computers in Spaceflight: The NASA Experience," in *Encyclopedia of Computer Science and Technology* 18, suppl. 3, ed. Allen Kent and James G. Williams (New York: Marcel Dekker, 1987), 40. Malvino and Leach explain, "Because of its inherent ruggedness, the core is a particularly useful logical element in applications where environmental extremes are experienced, for example, the temperature extremes and radiation exposure experienced by space vehicles." Malvino and Leach, *Digital Principles and Applications*, 331.

70. Eldon C. Hall, *MIT's Role in Project Apollo*, vol. 3 (Cambridge, MA: Charles Stark Draper Laboratory, 1972), 90–93.

71. As quoted in the video *Computer for Apollo*, dir. Russell Morash (Cambridge, MA: MIT/WGBH, 1965).

72. Tomayko, "Computers in Spaceflight," 38.

73. See David A. Mindell, *Digital Apollo: Human and Machine in Spaceflight* (Cambridge, MA: MIT Press, 2008). On the design and functioning of the guidance system, see also Paul E. Ceruzzi, *Beyond the Limits: Flight Enters the Computer Age* (Cambridge, MA: MIT Press, 1989), 210–212.

74. The rope core weaving process is described and demonstrated in detail in *Computer for Apollo*, throughout which the middle-aged memory weavers are referred to as "girls" by the Raytheon representative. This 1965 video also demonstrates the automatic wire wrap machine, which connected modules of rope memory via a punched-card system in a process remarkably similar to loom weaving with the Jacquard device.

75. As quoted in the video *Moon Machines: The Navigation Computer*, dir. Nick Davidson and Christopher Riley (London: DOX Productions/Discovery Science Channel, 2008).

76. Mindell, *Digital Apollo*, 155.

77. Jan Mazurek, *Making Microchips: Policy, Globalization, and Economic Restructuring in the Semiconductor Industry* (Cambridge, MA: MIT Press, 1999), 50–53; and David Kaplan, *The Silicon Boys and Their Valley of Dreams* (New York: William Morrow, 1999), 56–57. See also Christophe Lécuyer and David C. Brock, *Makers of the Microchip: A Documentary History of Fairchild Semiconductor* (Cambridge, MA: MIT Press, 2010).

78. Glenna Matthews, *Silicon Valley, Women, and the California Dream* (Palo Alto, CA: Stanford University Press, 2003), 230.

79. Shiprock dedication commemorative brochure, Fairchild Semiconductor publication XX-OO-0650-89, September 6, 1969, http://archive.computerhistory.org/resources/access/text/2014/07/102725169-05-01-acc.pdf (accessed January 7, 2017).

80. Shiprock plant manager Paul Driscoll, as quoted in Lisa Nakamura, "Indigenous Circuits: Navajo Women and the Racialization of Early Electronic Manufacture," *American Quarterly* 66, no. 4 (December 2014): 926.

81. Shiprock dedication commemorative brochure.

82. Dexter Ellis, "Age of Electronics Aids Economy of the Navajos," *Deseret News*, November 24, 1965, 6.

83. UPI, "Forty Armed Indians Seize a Building on Navajo Land," *New York Times*, February 25, 1975, 42; and Mike Cassidy, "What Went Wrong at Shiprock," *San Jose Mercury News*, May 7, 2000, 18.

84. See the genealogical chart by Don Hoefler, Harry Smallwood, and James E. Vincler, http://corphist.computerhistory.org/corphist/documents/doc-45ff3e214d9ea.pdf?PHPSESSID=89ad1d889a28ce5a1a26d8a9b6cf2d4b (accessed January 7, 2017);

and Christophe Lécuyer, *Making Silicon Valley: Innovation and the Growth of High Tech, 1930–1970* (Cambridge, MA: MIT Press, 2006).

85. Rogers and Larsen, *Silicon Valley Fever* (New York: Basic Books, 1984), 145; Matthews, *Silicon Valley, Women, and the California Dream,* 142. For more on labor conditions and related social issues in Silicon Valley, see Dennis Hayes, *Behind the Silicon Curtain: The Seductions of Work in a Lonely Era* (Boston: South End Press, 1989); and David Naguib Pellow and Lisa Sun-Hee Park, *The Silicon Valley of Dreams: Environmental Injustice, Immigrant Workers, and the High-Tech Global Economy* (New York: New York University Press, 2002).

86. Barbara Ehrenreich and Annette Fuentes, "Life on the Global Assembly Line," *Ms.* 9, no. 7 (January 1981): 71. See also Rachael Grossman, "Women's Place in the Integrated Circuit," *Radical America* 14, no. 1 (1980): 29–50.

87. Thomas R. Howell, *The Microelectronics Race: The Impact of Government Policy on International Competition* (Boulder, CO: Westview Press, 1988), 85. A similarly alarmist tune can be found in Fred Warshofsky, *The Chip War: The Battle for the World of Tomorrow* (New York: Charles Scribner's Sons, 1989), 8–16. For the figures on this shift, see Dan M. Khanna, *The Rise, Decline, and Renewal of Silicon Valley's High Technology Industry* (New York: Garland Publishing, 1997), 15–26.

88. See Allen J. Scott and David P. Angel, "The Global Assembly-Operations of US Semiconductor Firms: A Geographical Analysis," *Environment and Planning A* 20, no. 8 (1988): 1047–1067; Rosalinda Pineda-Ofreneo, "Women, Work and Micro-Electronics: Focus on the Philippines," in *Women, Work, and Computerization: Forming New Alliances*, ed. Kea Tijdens, Mary Jennings, Ina Wagner, and Margaret Weggelaar (Amsterdam: North-Holland, 1989), 317.

89. For an examination of the assembly work and its relationship to gender and globalization, see Teri L. Caraway, *Assembling Women: The Feminization of Global Manufacturing* (Ithaca, NY: ILR Press, 2007).

90. Simon Partner, *Assembled in Japan: Electrical Goods and the Making of the Japanese Consumer* (Berkeley: University of California Press, 1999), 231.

91. Ibid., 209.

92. It is significant that even when Asian electronics companies have opened plants in Europe, as Matsushita and NEC did in Britain the 1980s and 1990s, most of the shop floor jobs have gone to women. "The fact that these new employment opportunities are mostly confined to women has aroused some controversy in South Wales, where skilled men are losing their jobs in the coal and heavy industries." Mary Saso, *Women in the Japanese Workplace* (London: Hilary Shipman, 1990), 177.

93. Ray A. Killian, *The Working Woman: A Male Manager's View* (New York: American Management Association, 1971), 24.

94. Susan S. Green, "Silicon Valley's Women Workers: A Theoretical Analysis of Sex Segregation in the Electronics Industry Labor Market," in *Women, Men, and the International Division of Labor*, ed. June C. Nash and María Patricia Fernández-Kelly (Albany: State University of New York Press, 1983), 292.

95. Les Levidow, "Women Who Make the Chips," in *Women, Work, and Gender Relations in Developing Countries: A Global Perspective*, ed. Parvin Ghorayshi and Claire Bélanger (Westport, CT: Greenwood Press, 1996), 45–46. See also Caraway, *Assembling Women*.

96. M. Patricia Marchak, *The Integrated Circus: The New Right and the Restructuring of Global Markets* (Montreal: McGill-Queen's University Press, 1991), 143.

97. Pun Ngai, *Made in China: Women Factory Workers in a Global Workplace* (Durham, NC: Duke University Press, 2005), 149.

98. Diane Elson and Ruth Pearson, "'Nimble Fingers Make Cheap Workers': An Analysis of Women's Employment in Third World Export Manufacturing," *Feminist Review* 7 (1981): 93. For an extended consideration of skill and gender in the contemporary globalized context, see Caraway, *Assembling Women*, 48–55.

99. Michael Sharpston, "International Subcontracting," *World Development* 4, no. 4 (1976): 334.

100. Adam Hocherman, "Going It Alone, Part III: Inside the Factory Walls," TechCrunch, May 1, 2010, https://techcrunch.com/2010/05/01/going-it-alone-part-iii-inside-the-factory-walls/ (accessed January 7, 2017).

101. For a consideration of exploitative labor practices and iPhone assembly in particular, see Jon Agar, *Constant Touch: A Global History of the Mobile Phone*, 2nd ed. (London: Icon, [2004] 2013), 213–219.

102. Helen Grace, "iPhone Girl: Assembly, Assemblages, and Affect in the Life of an Image," in *Public Space, Media Space*, ed. Chris Berry, Janet Harbord, and Rachel O. Moore (New York: Palgrave Macmillan, 2013), 135–161. See also Aditya Chakrabortty, "The Woman Who Nearly Dies Making Your iPad," *The Guardian*, August 5, 2013, https://www.theguardian.com/commentisfree/2013/aug/05/woman-nearly-died-making-ipad (accessed January 7, 2017).

103. Seth Perlow, "On Production for Digital Culture: iPhone Girl, Electronics Assembly, and the Material Forms of Aspiration," *Convergence* 17, no. 3 (August 2011): 246.

104. Ibid., 248.

105. Jennifer Terry and Melodie Calvert, "Machines/Lives," in *Processed Lives: Gender and Technology in Everyday Life*, ed. Jennifer Terry and Melodie Calvert (New York: Routledge, 1997), 7. Indeed, as Susan Strasser notes, the invention of the sewing

machine in the nineteenth century not only changed the nature of home labor, but also brought industrial production into the domestic sphere as women sewed garments for clothing manufacturers out of their homes. Such piecework has played a part in the Asian electronics industry as well. Partner identifies the practice in over two thousand Japanese electronics companies in the late 1960s, where home-based workers would assemble transistors and transformers, wind coils, and solder components. Strasser, *Never Done*, 139; Partner, *Assembled in Japan*, 216.

106. Chun, *Programmed Visions*, 2–4.

107. Alan Sondheim, "Further Notes on Codework," *American Book Review* 22, no. 6 (2001): 1–2, as quoted in Geoff Cox and Alex McLean, *Speaking Code: Coding as Aesthetic and Political Expression* (Cambridge, MA: MIT Press, 2013), 39.

108. Agar, *Constant Touch*, 200–201.

## 2   Image Fabric

1. Sadie Plant, *Zeroes + Ones: Digital Women and the New Technoculture* (New York: Doubleday, 1997), 189.

2. Lev Manovich, *The Language of New Media* (Cambridge, MA: MIT Press, 2001), 100; Lev Manovich, *Software Takes Command* (New York: Bloomsbury, 2013), 33.

3. Martin Lister and others argue that the digital image is not a distinct entity, but rather a network practice. See *The Photographic Image in Digital Culture*, ed. Martin Lister, 2nd ed. (New York: Routledge, 2013).

4. Vannevar Bush, "As We May Think," *Atlantic Monthly* 176 (July 1945): 101–108.

5. Larry J. Schaaf, *Out of the Shadows: Herschel, Talbot, and the Invention of Photography* (New Haven: Yale University, 1992), 33.

6. Talbot accomplished many of his photographic discoveries and techniques in 1834 and 1835, only returning to work on his processes when the invention of photography was made public in 1839. Allan Bromley claims that Babbage's work on the Analytical Engine began when construction of the Difference Engine halted in late 1833, and much of the conception and design was in place by 1838. Allan G. Bromley, "Charles Babbage's Analytical Engine, 1838," *Annals of the History of Computing* 4, no. 3 (July 1982): 196–217.

7. See Charles Babbage to William Henry Fox Talbot, January 19, 1838, Fox Talbot Collection, British Library, LA38-3, http://foxtalbot.dmu.ac.uk/letters/ transcriptFreetext.php?keystring=babbage&keystring2=&keystring3=&year1=1800& year2=1877&pageNumber=23&pageTotal=51&referringPage=1 (accessed January 7, 2017); and an undated letter from Talbot to Babbage, Department of Manuscripts, British Library, Add MS 37201 f551, http://foxtalbot.dmu.ac.uk/letters/

transcriptFreetext.php?keystring=babbage&keystring2=engine&keystring3=&year1 =1800&year2=1877&pageNumber=0&pageTotal=5&referringPage=0 (accessed January 7, 2017).

8. William Henry Fox Talbot to Charles Babbage, May 10, 1839, Department of Manuscripts, British Library, Add MS 37191 f159, http://foxtalbot.dmu.ac.uk/letters/ transcriptName.php?bcode=Babb-C&pageNumber=12&pageTotal=20&referring Page=0 (accessed January 7, 2017); William Henry Fox Talbot to Constance Talbot, February 2, 1840, Fox Talbot Collection, British Library, LA(H)40-1, http://foxtalbot. dmu.ac.uk/letters/ftbh.php?docnum=04015 (accessed January 7, 2017).

9. Charles Babbage to William Henry Fox Talbot, February 26, 1844, Fox Talbot Collection, British Library, LA44-9, http://foxtalbot.dmu.ac.uk/letters/transcriptName .php?bcode=Babb-C&pageNumber=19&pageTotal=20&referringPage=0 (accessed January 7, 2017).

10. Steve Edwards argues that photography brought the "sketching and fancywork" that had been feminized in the nineteenth-century commercialization of the picturesque, back into a masculine register of "machines and culture." See Steve Edwards, *The Making of English Photography: Allegories* (University Park: Pennsylvania State University Press, 2006), 27–28.

11. See William Henry Fox Talbot to John Herschel, February 19, 1839, Royal Society, HS 17:284, http://foxtalbot.dmu.ac.uk/letters/transcriptDate.php?month =2&year=1839&pageNumber=25&pageTotal=40&referringPage=1 (accessed January 7, 2017).

12. Geoffrey Batchen, "Electricity Made Visible," in *New Media, Old Media: A History and Theory Reader*, ed. Wendy Hui Kyong Chun and Thomas Keenan (New York: Routledge, 2006), 30.

13. "Many thanks for your kind attention in sending me your Photogenic specimens. … They are very interesting especially the Lace one." David Brewster to William Henry Fox Talbot, February 12, 1839, National Media Museum, 1937-4832, http://foxtalbot.dmu.ac.uk/letters/transcriptDate.php?month=2&year=1839&pageN umber=19&pageTotal=40&referringPage=0 (accessed January 7, 2017). Talbot's mother, who advocated strongly for his invention, said: "I wish you would do some worked muslin & lace I sent you—the veracity of those is level with the meanest capacity & is consequently popular—people claim 'how natural is that bit of lace!'" Elisabeth Theresa Feilding to William Henry Fox Talbot, June 29, 1839, Fox Talbot Collection, British Library, LA40-002, http://foxtalbot.dmu.ac.uk/letters/transcript Date.php?month=6&year=1839&pageNumber=11&pageTotal=13&referringPage=0 (accessed January 7, 2017). Talbot exhibited ninety-three images, most being contact prints of lithographs, plants, and textiles. There were also twenty-one scenes made with a camera, all taken on and around his estate. William Henry Fox Talbot,

"Photogenic Drawings Exhibited in 1839," in *Henry Fox Talbot: Selected Texts and Bibliography*, ed. Mike Weaver (Oxford: Clio, 1992), 57–58.

14. Theresa Ann Maria Digby to William Henry Fox Talbot, April 13, 1839, Fox Talbot Collection, British Library, LA39-032, http://foxtalbot.dmu.ac.uk/letters/transcriptDate.php?month=4&year=1839&pageNumber=11&pageTotal=21&referringPage=0 (accessed January 7, 2017). For an illustrated example of Talbot's ribbon prints, see Schaaf, *Out of the Shadows*, 105. A contact print by Talbot of lace ribbon is reproduced in Gail Buckland, *Fox Talbot and the Invention of Photography* (Boston: David R. Godine, 1980), 122.

15. George Butler to William Henry Fox Talbot, May 31, 1839, Fox Talbot Collection, British Library, LA39-043, http://foxtalbot.dmu.ac.uk/letters/transcriptDate.php?month=5&year=1839&pageNumber=10&pageTotal=13&referringPage=0 (accessed January 7, 2017).

16. William Thomas Horner Strangways to William Henry Fox Talbot, March 7, 1840, Fox Talbot Collection, British Library, LA40-32, http://foxtalbot.dmu.ac.uk/letters/transcriptDate.php?month=3&year=1840&pageNumber=5&pageTotal=12&referringPage=0 (accessed January 7, 2017).

17. W. Hamish Fraser, "The Working Class," in *Glasgow, Volume II: 1830–1912*, ed. W. Hamish Fraser and Irene Mayer (Manchester: Manchester University Press, 1996), 109 and 304.

18. Douglas R. Nickel, "*Nature's Supernaturalism*: William Henry Fox Talbot and Botanical Illustration," in *Intersections: Lithography, Photography, and the Traditions of Printmaking*, ed. Kathleen Stewart Howe (Albuquerque: University of New Mexico Press, 1998), 19.

19. Geoffrey Batchen, *Each Wild Idea: Writing, Photography, History* (Cambridge, MA: MIT Press, 2002), 167.

20. Ibid., 169.

21. Talbot's insistence on the photograph as a "picture which makes itself," Steve Edwards has shown, mirrors descriptions of labor and textile production at the time that aimed for the absence of the hand in production. Edwards points out that it was the same ideology that guided Babbage's Analytical and Difference engines as calculating devices that would obviate the need for mathematical labor. See Edwards, *The Making of English Photography*, 31.

22. Larry J. Schaaf, "'The Caxton of Photography': Talbot's Etching of Light," in *William Henry Fox Talbot: Beyond Photography*, ed. Mirjam Brusius, Katrina Dean, and Chitra Ramalingam (New Haven: Yale University Press, 2013), 169–173. Sean Cubitt argues that this halftone grid is instrumental in establishing the raster image display format that has prevailed in later image technologies including cathode-ray tube television and LED digital screens. "Now the raster grid is normalized as the form

not just of the display, but in many instances of the signal being prepared for them," he claims. However, as with many other theorists of digital culture, Cubitt associates this with Cartesian grids originating in mathematical graphs as "one of the great signatures of modernity, from urban planning to modern art, and from the cartographers' longitude and latitude to the rows and columns of Microsoft Excel." Cubitt's geneaology effectively ties the raster grid to an exclusively phallocentric history, excluding the grid's much older and more ubiquitous application in textiles. See Sean Cubbitt, "LED Technology and the Shaping of Culture," in *Urban Screens Reader*, ed. Scott McQuire, Meredith Martin, and Sabine Niederer (Amsterdam: Institute of Network Cultures, 2009), 103.

23. Photo emulsion masks—essentially photographic negatives—have also been employed in printing chip surfaces, though it is an uncommon method. Keith Leaver, *Microelectronic Devices* (New York: John Wiley & Sons, 1989), 131.

24. Contact printing and the similar proximity printing process were common from the inception of the microchip through the 1970s. Since then, most chips have been produced through projection printing. Gene E. Fuller, "Optical Lithography," in *Handbook of Semiconductor Manufacturing Technology*, 2nd ed., ed. Robert Doering and Yoshio Nishi (Boca Raton: CRC Press, 2008), 18-1.

25. LG Electronics' 55EM960V OLED TV, for example, has a 140-centimeter (55-inch) screen display, but a depth of only four millimeters.

26. Anne Friedberg, *The Virtual Window: From Alberti to Microsoft* (Cambridge, MA: MIT Press, 2006), 1.

27. Jacques Derrida, *The Truth in Painting*, trans. Geoff Bennington and Ian McLeod (Chicago: University of Chicago Press, 1987), 60–61. Emphasis in original.

28. For a similar argument specific to cinema architecture and the screen, see Siegfried Kracauer, "Cult of Distraction: on Berlin's Picture Palaces," in Siegfried Kracauer, *The Mass Ornament: Weimar Essays*, ed. and trans. Thomas Y. Levin (Cambridge, MA: Harvard University Press, 1995), 323–328.

29. Derrida, *The Truth in Painting*, 57.

30. Stanley Cavell, *The World Viewed: Reflections on the Ontology of Film* (Cambridge, MA: Harvard University Press, 1979), 24–25.

31. This is similar to the production process of ikat fabrics, where thread is selectively dyed, sometimes with multiple colors, to produce pre-determined images and patterns that emerge only during the weaving process. See Jackie Battenfield, *Ikat Technique* (New York: Van Nostrand Reinhold, 1978); and Manisha Iyer, "Les Textiles Patola de Patan (Inde), XVIIe et XXIe Siècles: Techniques, Patrimoine, Mémoire" (PhD diss., Université Paris 1—Panthéon-Sorbonne, 2015).

32. For more on Fairchild's early adoption of photolithographic methods developed at the Diamond Ordnance Fuze Laboratory and Bell Laboratories, see Christophe

Lecuyer and David C. Brock, *Makers of the Microchip: A Documentary History of Fairchild Semiconductor* (Cambridge, MA: MIT Press, 2010), 19–20.

33. Edmond Couchot, "The Ordered Mosaic, or the Screen Overtaken by Computation," trans. Stephen Monteiro, in *The Screen Media Reader*, ed. Stephen Monteiro (New York: Bloomsbury Academic, 2017), 134.

34. Geoffrey Batchen, *Burning with Desire: The Conception of Photography* (Cambridge, MA: MIT Press, 1997), 216. For more on these debates, see Martin Hand, *Ubiquitous Photography* (Cambridge: Polity, 2012), 59–95.

35. William J. Mitchell, *The Reconfigured Eye: Visual Truth in the Post-Photographic Era* (Cambridge, MA: MIT Press, 1992), 4–5.

36. Couchot, "The Ordered Mosaic," 137.

37. Batchen, *Each Wild Idea*, 179.

38. Mitchell nominates "digital image," while Martin Lister suggests "photographic image" as "more generic and less causal than 'photograph.'" See Mitchell, *The Reconfigured Eye*, 49, and Martin Lister, "Introduction," in *The Photographic Image in Digital Culture*, ed. Martin Lister (London: Routledge, 1995), 3.

39. Lister, "Introduction," 2; and Kevin Robins, "Will the Image Move Us Still?," in *The Photographic Image in Digital Culture*, ed. Martin Lister (London: Routledge, 1995), 30–31.

40. Mitchell, *The Reconfigured Eye*, 24 and 28.

41. John Tagg, *The Burden of Representation: Essays on Photographies and Histories* (Amherst: University of Massachusetts Press, 1988), 2–4. Emphasis in original. John Roberts asserts: "In essence, what the new imaging techniques produce is not the death of the 'truth' of the photograph, but the cultural displacement of the indexicality of the photograph as an *automatic truth-effect*." John Roberts, *The Art of Interruption: Realism, Photography, and the Everyday* (Manchester: Manchester University Press, 1998), 221. Emphasis in original.

42. Fred Ritchin, *After Photography* (New York: W. W. Norton, 2009), 31. For an interpretation of this passage, see Sarah Kember, *Virtual Anxiety: Photography, New Technologies, and Subjectivity* (Manchester: Manchester University Press, 1998), 28–29.

43. "Because the computer works with small bits of the image, alterations can be totally *seamless* and wholly realistic." See Gloria S. McDarrah, Fred W. McDarrah, and Timothy S. McDarrah, *The Photography Encyclopedia* (New York: Schirmer Books, 1999), 92; Fred Ritchin, "Photojournalism in the Age of Computers," in *The Critical Image: Essays on Contemporary Photography*, ed. Carol Squiers (Seattle: Bay Press, 1990), 29.

44. Mitchell, *The Reconfigured Eye*, 7.

45. The JPEG format is known for image-quality loss when compressed, producing artifacts and a patchwork effect, while PNG is slower but maintains its quality when compressed.

46. Ritchin, *After Photography*, 141.

47. Dennis R. Cohen and Erica Sadun, *Mac Digital Photography* (Alameda, CA: SYBEX, 2003), 115.

48. Chun-Wei Hsieh, Tse Cheng, Cheng-Chin Chiang, and Shu-Cheng Huang, "Intelligent Stitcher for Panoramic Image-Based Virtual Worlds," U.S. patent 6011558 A, filed September 23, 1997, and issued January 4, 2000.

49. This process is also known as feathering. See Ken Milburn, *Digital Photography: Expert Techniques* (Sebastopol, CA: O'Reilly, 2004), 133.

50. The Butterick Publishing Company, *The New Dressmaker* (New York: The Butterick Publishing Company, 1921), 146–147.

51. Joseph A. Ippolito, *Understanding Digital Photography* (New York: Thomson/ Delmar Learning, 2003), 180.

52. "Combine images to create a panoramic photograph. … Click Browse [sic] to add images that you want to combine." See dialog boxes illustrated in Adobe, *Adobe Photoshop Elements 5.0 Classroom in a Book* (Berkeley, CA: Adobe Press, 2007), 222– 226.

53. Molly Joss, *How to Do Everything with Photoshop Elements* (Berkeley, CA: Osborne/ McGraw-Hill, 2001), 129. See also Deke McClelland and Galen Fott, *Photoshop Elements for Dummies* (New York: Hungry Minds, 2001), 320.

54. Mike Woolridge, *Teach Yourself Visually Photoshop Elements 2* (New York: Wiley, 2003), 262.

55. McClelland and Fott, *Photoshop Elements for Dummies*, 320.

56. Milburn, *Digital Photography*, 95.

57. Alexei A. Efros and William T. Freeman, "Image Quilting for Texture Synthesis and Transfer," *SIGGRAPH '01* (2001): 341.

58. See "Google Maps API: Street View Service," Google, https://developers.google .com/maps/documentation/javascript/streetview (accessed January 7, 2017).

59. For background into the development of packet switching in data networks, see Lawrence G. Roberts, "The Evolution of Packet Switching," *Proceedings of the IEEE* 66, no. 11 (November 1978): 1307–1313.

60. Daniel Rubinstein and Katrina Sluis, "The Digital Image in Photographic Culture: Algorithmic Photography and the Crisis of Representation," in *The Photographic Image in Digital Culture*, ed. Martin Lister, 2nd ed. (New York: Routledge, 2013), 30.

61. Julia H. Farago, Hugh E. Williams, James E. Walsh, Nicholas A. Whyte, et al., "Object Search UI and Dragging Object Results," U.S. patent 7664739 B2, filed February 14, 2006, and issued February 16, 2010. See also Hugh Williams, "Ideas and Invention (and the Story of Bing's Image Search)," Hugh E. Williams, https://hughewilliams.com/2012/03/06/ideas-and-invention-and-the-story-of-bings-image-search/ (accessed January 7, 2017).

62. Yogev Ahuvia, "Infinite Scrolling: Let's Get to the Bottom of This," Smashing Magazine, May 3, 2013, https://www.smashingmagazine.com/2013/05/03/infinite-scrolling-lets-get-to-the-bottom-of-this/ (accessed January 7, 2017).

63. Ibid.; see also Mike Takahashi "Should You Use Infinite Scroll Instead of Pagination to Load More Content?," Taka Designs, November 29, 2010, http://www.takadesigns.com/blog/2010/11/29/should-you-use-infinite-scroll-instead-of-pagination-to-load-more-content/ (accessed January 7, 2017).

64. Dmitry Fadeyev, "When Infinite Scroll Doesn't Work," Usability Post, January 7, 2013, http://usabilitypost.com/2013/01/07/when-infinite-scroll-doesnt-work/ (accessed January 7, 2017). Emphasis in original.

65. Since digital culture's relationship to quilting as a networked social practice is taken up in chapter 3, consideration here is restricted to quilting's formal visual qualities.

66. Hettie Risinger, *Innovative Machine Quilting* (New York: Sterling Publishing, 1980), 22.

67. André Gunthert, "The Conversational Image: New Uses of Digital Photography," trans. Fatima Aziz, *Études photographiques*, no. 31 (Spring 2014), https://etudesphotographiques.revues.org/3546 (accessed January 7, 2017).

68. Rubinstein and Sluis, "The Digital Image in Photographic Culture," 31.

69. Carleton L. Safford and Robert Bishop, *America's Quilts and Coverlets* (New York: Bonanza Books, 1980), 145.

70. It would appear that is not strictly the case. A search of a well-known monument at sunset, for example, will produce results of the same monument in similar colors.

71. Rubinstein and Sluis, "The Digital Image in Photographic Culture," 33.

72. Matthew Panzarino, "Instagram's Latest App Update Fixed My Biggest Pet Peeve, and People Hate It," The Next Web, December 18, 2012, http://thenextweb.com/apps/2012/12/18/instagrams-latest-app-update-fixed-my-biggest-pet-peeve-and-people-hate-it/ (accessed January 7, 2017).

73. Liz Stinson, "Instagram Ends the Tyranny of the Square," *Wired*, August 27, 2015, https://www.wired.com/2015/08/instagram-says-goodbye-square-photos/ (accessed January 7, 2017).

74. Josh Constine, "Snapchat Memories Is a Searchable Replacement for Your Camera Roll," Techcrunch, July 6, 2016, https://techcrunch.com/2016/07/06/snapchat-memories/ (accessed January 7, 2017).

75. Safford and Bishop, *America's Quilts and Coverlets*, 145.

76. Seth Fiegerman, "Snapchat Introduces SnapKidz for Users under 13," Mashable, June 24, 2013, http://mashable.com/2013/06/24/snapchat-kidz/#JiSzLrnyDgqO (accessed January 7, 2017).

77. Martin Lister, "Introduction," in *The Photographic Image in Digital Culture*, ed. Martin Lister, 2nd ed. (New York: Routledge, 2013), 8.

## 3   Piecework

1. Ben Shneiderman, *Leonardo's Laptop: Human Needs and the New Computing Technologies* (Cambridge, MA: MIT Press, 2002), 2.

2. Wendy Hui Kyong Chun, *Updating to Remain the Same: Habitual New Media* (Cambridge, MA: MIT Press, 2016).

3. Claire Zillman, "Facebook Is Now More Valuable than Exxon," *Fortune*, February 1, 2016, http://fortune.com/2016/02/01/facebook-value-exxon/ (accessed January 7, 2017).

4. Even when such assembly is minimal, as with browsing and scrolling, the user's path and choices through this material leave new wrinkles in the algorithms that manage it, shaping when and how it will appear in the future.

5. More recently, when the AIDS crisis struck America in the 1980s, activists turned to the production of a large-scale quilt, the AIDS Memorial Quilt, eventually comprising over 48,000 panels. Although representing all affected by the disease, it has also served as a visual and material representation of the particularly large toll the virus has taken among African Americans and gay populations. See Charles E. Morris III, *Remembering the AIDS Quilt* (East Lansing: Michigan State University, 2011).

6. Even sites that explicitly trade in paid online labor, such as Amazon's Mechanical Turk, will frame the process in ways that suggest it may function as leisure-time entertainment on par with casual gaming and social media interactivity. See Geoff Cox and Alex McLean, *Speaking Code: Coding as Aesthetic and Political Expression* (Cambridge, MA: MIT Press, 2013), 48–49.

7. David Weinberger, *Small Pieces Loosely Joined: A Unified Theory of the Web* (Cambridge, MA: Perseus Publishing 2002), x.

8. Chris Messina is credited for launching hashtags on Twitter with an August 23, 2007, tweet "how do you feel about using # (pound) for groups. As in #barcamp [msg]?"

9. Theodor H. Nelson, *Literary Machines*, vers. 87/1 (Swarthmore, PA: Theodor H. Nelson, 1987), 0/2.

10. George P. Landow, *Hypertext: The Convergence of Contemporary Critical Theory and Technology* (Baltimore: Johns Hopkins University Press, 1992), 4.

11. Jenny Sundén, "What If Frankenstein('s Monster) Was a Girl?: Reproduction and Subjectivity in the Digital Age," in *Bits of Life*, ed. Anneke M. Smelik and Nina Lykke (Seattle: University of Washington Press, 2008), 158.

12. Sheri Pocilujko, "10 Reasons Women Like Casual Games: Why Casual Games and Female Gamers Go Together," *Casual Connect Magazine*, Fall 2006, http://www.casualconnect.org/content/gamedesign/pocilujko-ten.html (accessed January 7, 2017).

13. Alison Harvey, *Gender, Age, and Digital Games in the Domestic Context* (New York: Routledge, 2015), 33, 131.

14. Aubrey Anable, "Casual Games, Time Management, and the Work of Affect," *Ada* 2 (2013), http://adanewmedia.org/2013/06/issue2-anable/ (accessed January 7, 2017).

15. Sven Grundberg and Jens Hansegard "Women Now Make Up Almost Half of Gamers," *Wall Street Journal*, August 20, 2014, http://www.wsj.com/articles/gaming-no-longer-a-mans-world-1408464249 (accessed January 7, 2017); Keith Stuart, "UK Gamers: More Women Play Games than Men, Report Finds," *The Guardian*, September 17, 2014, https://www.theguardian.com/technology/2014/sep/17/women-video-games-iab (accessed January 7, 2017).

16. Harvey, *Gender, Age, and Digital Games in the Domestic Context*, 32–33. For the historically male—and misogynist—orientation of gamer culture, see also Adrienne Shaw, *Gaming at the Edge: Sexuality and Gender at the Margins of Gamer Culture* (Minneapolis: University of Minnesota, 2014); and Mia Consalvo, "Confronting Toxic Gamer Culture: A Challenge for Feminist Game Studies Scholars," *Ada* 1 (2012), http://adanewmedia.org/2012/11/issue1-consalvo/ (accessed January 7, 2017).

17. Kyle Stock, "Highlights from the Candy Crush IPO Filing: 500 Million Downloads and Counting," Bloomberg, February 18, 2014, https://www.bloomberg.com/news/articles/2014-02-18/king-digitals-ipo-filing-shows-500-million-candy-crush-downloads (accessed January 7, 2017).

18. Aubrey Anable, "Casual Games, Time Management, and the Work of Affect."

19. Andrew Webster, "Half a Billion People Have Installed 'Candy Crush Saga,'" The Verge, November 15, 2013, http://www.theverge.com/2013/11/15/5107794/candy-crush-saga-500-million-downloads (accessed January 7, 2017).

20. The tracing line also exists for other genres, such as word-puzzle apps including WordWhizzle by Apprope.

21. Gordon Calleja, *In-Game: From Immersion to Incorporation* (Cambridge, MA: MIT Press, 2011), 27.

22. Brian Janosch, "Patrick Moberg, Dots God: How High Art and Bill Cosby Inspired the Hottest Mobile Game," Neat, May 13, 2013, http://www.cultivatedwit.com/patrick-moberg-dots-god-how-high-art-bill-cosby-inspired-the-hottest-mobile-game/ (accessed January 7, 2017).

23. Matthew Lynley, "Behind the Hit Game 'Dots,'" *Wall Street Journal*, May 7, 2013, http://blogs.wsj.com/digits/2013/05/07/behind-the-hit-game-dots/ (accessed January 7, 2017).

24. Seth Fiegerman, "Nearly 20 Million Downloads Later, Dots Looks to Build a Gaming Franchise," Mashable, April 30, 2014, http://mashable.com/2014/04/30/dots-profile/#b15Ke.oDbGq6 (accessed January 7, 2017). Emma Westecott has argued for another approach to game design that involves craft culture at an ideological level. Studying independent game production outside commercial, mass-market models, she identifies opposing possibilities for empowerment and exploitation in its design practices, particularly concerning the historically marginalized role of women in gaming. See Emma Westecott, "Independent Game Development as Craft," *Loading … The Journal of the Canadian Game Studies Association* 7, no. 11 (2013): 78–91.

25. The Jeff Rubin Show, " 'Dots' Creator Patrick Moberg," April 28, 2014, https://soundcloud.com/jeffrubinjeffrubin/118-dots-creator-patrick (accessed January 7, 2017).

26. Matthew Fuller and Andrew Goffey, *Evil Media* (Cambridge, MA: MIT Press, 2012), 52.

27. Not surprisingly, then, sites explicitly oriented toward professional goals, such as LinkedIn, emphasize network over community.

28. Robert Shaw, *American Quilts: The Democratic Art, 1780–2007* (New York: Sterling, 2009), 7.

29. Whole cloth quilts use a single piece of fabric, upon which stitched patterns produce the design and texture. Appliqué quilts involve hemming smaller pieces of fabric onto a larger, underlying fabric.

30. For numerous examples, from Central Asia to the Middle East and South America, see Caroline Crabtree and Christine Shaw, *Quilting, Patchwork and Appliqué: A World Guide* (London: Thames & Hudson, 2007).

31. Barbara Brackman, "American Adaptation: Block-Style Quilts," in *American Quilts in the Modern Age, 1870–1940: The International Quilt Study Center Collections,*

ed. Marin F. Hanson and Patricia Cox Crews (Lincoln: University of Nebraska Press, 2009), 22–23.

32. O. W. Scott, "Aunt Bina's Quilt," reprinted in *A Patchwork of Pieces: An Anthology of Early Quilt Stories, 1845–1949*, ed. Cuesta Ray Benberry and Carol Pinney Crabb (Paducah, KY: American Quilter's Society, 1993), 170.

33. Ibid., 171.

34. Ibid., 176.

35. Cuesta Ray Benberry and Carol Pinney Crabb, eds., *A Patchwork of Pieces: An Anthology of Early Quilt Stories, 1845–1949* (Paducah, KY: American Quilter's Society, 1993), 12.

36. For other examples of narrative and communication through textile crafts across cultures, and their potential relation to digital practices, see Jack Bratich, "The Digital Touch: Craft-Work as Immaterial Labour and Ontological Accumulation," *Ephemera* 10, no. 3–4 (2010): 306.

37. Ruth Finley, *Old Patchwork Quilts* (Philadelphia: J. B. Lippincott, 1929), 33.

38. Bratich, "The Digital Touch," 307.

39. Marguerite Ickis, *The Standard Book of Quilt Making and Collecting* (New York: Dover, 1959), vii.

40. Lucy R. Lippard, "Up, Down, and Across: A New Frame for New Quilts," in *The Artist and The Quilt*, ed. Charlotte Robinson (New York: Alfred A. Knopf, 1983), 32.

41. Tony Blackshaw, *Leisure* (New York: Routledge, 2010), x–xi.

42. In a 2015 Pew study, 59 percent of smartphone owners said they used their apps "several times a day" and 27 percent said they used them "continuously" (this jumped to 43 percent among those eighteen to twenty-nine years old. Lee Rainie and Kathryn Zickuhr, "Americans' Views on Mobile Etiquette," Pew Research Center, August 26, 2015, http://www.pewinternet.org/2015/08/26/americans-views-on -mobile-etiquette/ (accessed January 7, 2017).

43. David Rowe, "Leisure, Mass Communications and Media," in *A Handbook of Leisure Studies*, ed. Chris Rojek, Susan M. Shaw, and A. J. Veal (New York: Palgrave Macmillan, 2006), 330.

44. Ibid., 329.

45. In "The Concept of Leisure," philosopher Cyril Barrett describes the complexity of separating leisure from work time by explaining that compulsory chores in leisure time must be described as work while "knitting or playing cards, if that is what one wants to do, is leisure, even if done during office hours or in work time." Of course, the two have little in common, since knitting is a hand skill that produces clothes

while card games are strategy tasks—depending on the game—often tied to gambling. And while someone may desire to knit while at another paid task, who's to say the garments produced aren't being made out of necessity? Cyril Barrett, "The Concept of Leisure: Idea and Ideal," in *The Philosophy of Leisure*, ed. Tom Winnifrith and Cyril Barrett (Basingstoke: Macmillan, 1989), 11.

46. Susan Strasser, *Never Done: A History of American Housework* (New York: Pantheon, 1982), 185.

47. Ruth Schwartz Cowan, *More Work for Mother: The Ironies of Household Technology from the Open Hearth to the Microwave* (New York: Basic Books, 1983), 63–65. Strasser notes that the sewing machine also introduced increased industrial production in the home through piecework for manufacturers. See Strasser, *Never Done*, 139.

48. Melissa Gregg, *Work's Intimacy* (Cambridge: Polity, 2011), 3.

49. David Staples, "Women's Work and the Ambivalent Gift of Entropy," in *The Affective Turn: Theorizing the Social*, ed. Patricia Ticineto Clough and Jean O'Malley Halley (Durham, NC: Duke University Press, 2007), 125–126.

50. Maurizio Lazzarato, "Immaterial Labor," in *Marxism Beyond Marxism*, ed. Saree Makdisi, Cesare Casarino, and Rebecca E. Karl (London: Routledge, 1996), 133.

51. Jan L. Harris and Paul A. Taylor, *Digital Matters: Theory and Culture of the Matrix* (New York: Routledge, 2005), 186.

52. Cox and McLean, *Speaking Code*, 49.

53. David Staples, "Women's Work and the Ambivalent Gift of Entropy," in *The Affective Turn*, ed. Patricia Ticineto Clough and Jean O'Malley Halley (Durham, NC: Duke University Press, 2007), 125.

54. Tiziana Terranova, *Network Culture: Politics for the Information Age* (London: Pluto Press, 2004), 74.

55. See *Homeworkers in Global Perspective: Invisible No More*, ed. Eileen Boris and Elisabeth Prügl (New York: Routledge, 1996); Ping-Chun Hsiung, *Living Rooms as Factories: Class, Gender, and the Satellite Factory System in Taiwan*; Charles B. Hennon and Suzanne Loker, "Gender and Home-Based Employment in a Global Economy," in *Gender and Home-Based Employment*, ed. Charles B. Hennon, Suzanne Loker, and Rosemary Walker (Westport, CT: Auburn House, 2000), 17–43.

56. Eileen Boris, *Home to Work: Motherhood and the Politics of Industrial Homework in the United States* (Cambridge: Cambridge University Press, 1994), 10.

57. Susan M. Strasser, "An Enlarged Human Existence? Technology and Household Work in Nineteenth-Century America," in *Women and Household Labor*, ed. Sarah Fenstermaker Berk (Beverly Hills, CA: SAGE, 1980), 45.

58. Ibid., 46; see also Sheila Allen and Carol Wolkowitz, "Homeworking and the Control of Women's Work," in *Waged Work: A Reader*, ed. Feminist Review (London: Virago Press, 1986), 238–264.

59. Cynthia R. Daniels, "Between Home and Factory," in *Homework: Historical and Contemporary Perspectives on Paid Labor at Home*, ed. Eileen Boris and Cynthia R. Daniels (Urbana: University of Illinois Press, 1989), 14–15.

60. Jamie Faricellia Dangler, "Electronics Subassemblers in Central New York," in *Homework: Historical and Contemporary Perspectives on Paid Labor at Home*, ed. Eileen Boris and Cynthia R. Daniels (Urbana: University of Illinois Press, 1989), 147–164.

61. Ibid., 153.

62. Ibid., 152. In that same period, 20 percent of New York City's 250,000 garment workers were working illegally in sweatshops and as homeworkers. See Boris, *Home to Work*, 342.

63. Laura C. Johnson, *The Seam Allowance: Industrial Home Sewing in Canada* (Toronto: Women's Educational Press, 1982), 9.

64. Haraway's essay first appeared as "Manifesto for Cyborgs," but has been published and reprinted subsequently as "A Cyborg Manifesto." See Donna Haraway, "Manifesto for Cyborgs: Science, Technology, and Socialist Feminism in the 1980s," *Socialist Review*, no. 80 (1985): 65–108; and Donna J. Haraway, "A Cyborg Manifesto: Science, Technology, and Socialist-Feminism in the Late Twentieth Century," chap. 8 in *Simians, Cyborgs and Women: The Reinvention of Nature* (New York: Routledge, 1991), 149–181.

65. Haraway, "A Cyborg Manifesto," 166.

66. Ibid., 170.

67. Martha Rosler, "Image Simulations, Computer Manipulations: Some Considerations," in *Photography after Photography: Memory and Representation in the Digital Age*, ed. Hubertus von Amelunxen, Stefan Iglhaut, Florian Rötzer, and Alexis Cassel (Amsterdam: G+B Arts, 1996), 50.

68. Nicole B. Ellison, *Telework and Social Change: How Technology Is Reshaping the Boundaries between Home and Work* (Westport, CT: Praeger, 2004), 116.

69. Ibid., 120. Emphasis in original.

70. Ned Rossiter, *Organized Networks: Media Theory, Creative Labour, New Institutions* (Amsterdam: Institute of Network Cultures, 2006), 14–15.

71. Jan van Dijk, *The Network Society*, 3rd ed. (London: SAGE, 2012), 181.

72. Terranova, *Network Culture*, 74.

73. Ibid., 77.

74. Ibid., 91–92.

75. Bratich, "The Digital Touch," 304.

## 4   Domestic Disturbances

1. Device accessories like Logitech's AnyAngle stands perpetuate this idea of unlimited body-device orientations. See Logitech, "The New Logitech AnyAngle Brings Flexibility and Protection to the iPad Air 2 and iPad mini," http://news.logitech .com/press-release/consumer-products/new-logitech-anyangle-brings-flexibility -and-protection-ipad-air-2-a (accessed January 7, 2017).

2. Mindy McAdams, "Gender without Bodies," *CMC Magazine* 3, no. 3 (March 1996), http://www.december.com/cmc/mag/1996/mar/mcadams.html (accessed January 7, 2017).

3. Ken Pimentel and Kevin Teixeira, *Virtual Reality: Through the New Looking Glass* (New York: Intel/Windcrest/McGraw Hill, 1993), 7–8.

4. Kees Overbeeke, Tom Djajadiningrat, Caroline Hummels, Stephan Wensveen, and Joep Frens, "Let's Make Things Engaging," in *Funology: From Usability to Enjoyment*, ed. Mark A. Blythe, Kees Overbeeke, Andrew F. Monk, and Peter C. Wright (Dordrecht: Kluwer Academic Publishers, 2003), 7 and 16.

5. Donald A. Norman, "Why Interfaces Don't Work," in *The Art of Human–Computer Interface Design*, ed. Brenda Laurel (Reading, MA: Addison-Wesley, 1990), 209.

6. See Kimberly A. Neuendorf and Evan A. Lieberman, "Film: The Original Immersive Medium," in *Immersed in Media: Telepresence in Everyday Life*, ed. Cheryl Campanella Bracken and Paul D. Skalski (New York: Routledge, 2010), 9–38.

7. See Wanda Strauven, "The Observer's Dilemma: To Touch or Not to Touch," in *Media Archaeology: Approaches, Applications, and Implications*, ed. Erkki Huhtamo and Jussi Parikka: 148–163. Strauven even claims an "archaeology of the 'touch screen'" grounded in early cinema systems such as the Mutoscope, which required the viewer to crank a handle to set and keep images in motion. See ibid., 155–158.

8. Erkki Huhtamo, "Screenology; or, Media Archaeology of the Screen," in *The Screen Media Reader: Culture, Theory, Practice*, ed. Stephen Monteiro (New York: Bloomsbury Academic, 2017), 108.

9. Designers speak of "hand- and finger-based occlusion." It is compensated for through interface features such as keyboard offset controls for selecting letters.

10. Teresa de Lauretis, *Technologies of Gender: Essays on Theory, Film, and Fiction* (Bloomington: Indiana University Press, 1987), 13.

11. Jacques Rancière, *The Politics of Aesthetics: The Distribution of the Sensible*, trans. Gabriel Rockhill (London: Continuum, 2004), 13.

12. Constance Classen, "Feminine Tactics: Crafting an Alternative Aesthetics in the Eighteenth and Nineteenth Centuries," in *The Book of Touch*, ed. Constance Classen (Oxford: Berg, 2005), 228.

13. See Nicholas Mirzoeff, *The Right to Look: A Counterhistory of Visuality* (Durham, NC: Duke University Press, 2011).

14. Elizabeth Grosz, *Volatile Bodies: Toward a Corporeal Feminism* (Bloomington: Indiana University Press, 1994), 3–4.

15. Maurice Merleau-Ponty, *The Phenomenology of Perception*, trans. Colin Smith (London: Routledge, [1962] 2002), 369. Interestingly, when explaining tactile perception, Merleau-Ponty relies on the example of handling textiles. See ibid., 368–369.

16. Elizabeth Grosz ties Merleau-Ponty's ideas on tactility to gender distinctions in *Volatile Bodies*, 98–107. See also Heidi J. Nast and Audrey Kobayashi, "Re-corporealizing Vision," in *Bodyspace: Destabilizing Geographies of Gender and Sexuality*, ed. Nancy Duncan (London: Routledge, 1996), 75–93.

17. Christian Metz, "The Imaginary Signifier," *Screen* 16, no. 2 (Summer 1975): 60.

18. Mirzoeff, *The Right to Look*, 3.

19. See, for example, *The Feminism and Visual Culture Reader*, ed. Amelia Jones (New York: Routledge, 2003).

20. See Laura Mulvey, "Visual Pleasure and Narrative Cinema," *Screen* 16, no. 3 (Autumn 1975): 6–18; Mary Ann Doane, "Film and Masquerade: Theorising the Female Spectator," *Screen* 23, no. 3–4 (1982): 74–88; Stephen Heath and Teresa de Lauretis, eds., *The Cinematic Apparatus* (New York: St. Martin's Press, 1980).

21. AMC CEO Adam Aron explained: "When you tell a 22-year-old to turn off the phone, don't ruin the movie, they hear please cut off your left arm above the elbow. You can't tell a 22-year-old to turn off their cellphone. That's not how they live their life." Brent Lang, "AMC Entertainment CEO Open to Allowing Texting in Some Theaters," *Variety*, April 13, 2016. See also AMC Theaters, "No Texting at AMC," Twitter, April 15, 2016, https://twitter.com/amctheatres/status/720972338699702272 (accessed January 7, 2017).

22. Constance Classen, *The Color of Angels: Cosmology, Gender and the Aesthetic Imagination* (New York: Routledge, 1998), 1.

23. Donna J. Haraway, "Situated Knowledges: The Science Question in Feminism and the Privilege of Partial Perspective," chap. 9 in *Simians, Cyborgs and Women: The Reinvention of Nature* (New York: Routledge, 1991), 189.

24. Allucquère Rosanne Stone, *The War of Desire and Technology at the Close of the Mechanical Age* (Cambrudge, MA: MIT Press, 1995); and Allucquère Rosanne Stone, "Will the Real Body Please Stand Up? Boundary Stories about Virtual Cultures," in *Cyberspace: First Steps*, ed. Michael Benedickt (Cambridge, MA: MIT Press, 1994), 81–118.

25. Priska Gisler, "Does Gender Still Matter? Bodily Functions in Cyberspace: A Feminist Approach," in *Women, Work and Computerization: Spinning a Web from Past to Future, Proceedings of the 6th International IFIP Conference, Bonn, Germany, May 24–27, 1997*, ed. A. Frances Grundy, Doris Köhler, Veronika Oechtering, and Ulrike Petersen, (New York: Springer, 1997), 219 and 220.

26. Anita Greenhill, "...Virtually There: The Social Construction of Computer Mediated Identity," Spaceless.com, http://www.spaceless.com/papers/13.htm (accessed January 7, 2017).

27. Charles W. Griffith, "Marksmanship Practicing Means," U.S. patent 2007082 A. filed April 12, 1934, and issued July 2, 1935. Chester Braselton filed a patent application for a television photo-cell pickup system in 1936. Chester Braselton, "Photo-Cell Pickup System," U.S. patent 2193789, filed January 20, 1936, and issued March 19, 1940.

28. "Seeburg Ray-o-Lite Presented after 20-Month Test," *Atomic Age* (January 1937): 243.

29. Lincoln Laboratory technicians initially had tested joysticks as selection devices, a choice that would have located the hand controls outside the frame of the screen, but light guns allowed military personnel to select and deselect screen objects more quickly. For discussion of the joystick and light gun, see "Bi-weekly, Project 6673," Memorandum M-2074, Electronic Computer Division, Servomechanisms Laboratory, Massachusetts Institute of Technology (October 27, 1950), 4. For details on the light gun's construction, see "Bi-weekly, Project 6673," Memorandum M-2075, Electronic Computer Division, Servomechanisms Laboratory, Massachusetts Institute of Technology (November 11, 1950), 9.

30. The same system would become commercially available as the Magnavox Odyssey Shooting Gallery in 1972, part of the first commercial home video game console.

31. SAGE was a continental surveillance system (the largest of its kind at the time), representing a scale of technological visualization that would only be superseded by the supra-stratospheric perspective of artificial satellites in the late 1950s. See Kent C. Redmond and Thomas M. Smith, *From Whirlwind to MITRE: The R&D Story of the SAGE Air Defense Computer* (Cambridge, MA: MIT Press, 2000).

32. Surprisingly, even Wendy Chun does not note this discrepancy in her analysis of the SAGE interface in *Programmed Visions*. She refers to the gun as a "light pen," although the book includes an illustration of the pistol grip device in use by

servicemen. See Wendy Hui Kyong Chun, *Programmed Visions: Software and Memory* (Cambridge, MA: MIT Press, 2011), 60–61.

33. C. R. Wieser, "Cape Cod System and Demonstration," Memorandum VI, L-86, Lincoln Laboratory, Massachusetts Institute of Technology (March 13, 1953). It further explains: "The light gun is a photocell device which is placed over the desired blip on the display scope and then sends a pulse into the computer to indicate to the computer that action (for example, 'start tracking') is to be taken on that particular aircraft. ... The human beings make decisions and improvise while the computer handles routine tasks under their supervision."

34. Benj Edwards, "The Never-Before-Told Story of the World's First Computer Art (It's a Sexy Dame)," *The Atlantic*, January 24, 2013, http://www.theatlantic.com/technology/archive/2013/01/the-never-before-told-story-of-the-worlds-first-computer-art-its-a-sexy-dame/267439/ (accessed January 7, 2017).

35. "SAGE A/N FSQ-7," http://www.smecc.org/sage_a_n_fsq-7.htm (accessed January 7, 2017).

36. For a consideration of SAGE's contributions to digital gaming culture, see Patrick Crogan, *Gameplay Mode: War, Simulation, and Technoculture* (Minneapolis: University of Minnesota Press, 2011), 6–12. Sadly, Crogan makes no mention of the Girley programs.

37. As Patrick Crogan explains in his study of the tightly interlinked history of military technology and the video game industry: "The military technoscientific legacy forged in the face of total war and the nuclear age inaugurated by the cold war ... inhabit[s] the technological lineages of digital computing, visual displays and interactivity, virtual space simulation, and software development." Crogan, *Gameplay Mode*, xii.

38. Branden Hookway, *Interface* (Cambridge, MA: MIT Press, 2014), 150.

39. Ivan Edward Sutherland, "Sketchpad: A Man-Machine Graphical Communication System" (PhD diss., MIT, 1963), 57–60. Unfortunately, many subsequent descriptions of Sketchpad mistakenly claim that the pen has to touch the screen to function, an error presumably derived from the title Sutherland chose for his program.

40. In the early development of mobile touchscreen devices, designers considered the possibility of using a crosshair as a screen cursor that would allow the finger to accurately select data on the screen. See Wilbert O. Galitz, *The Essential Guide to User Interface Design* (New York: John Wiley & Sons, 1997), 309.

41. Sutherland, "Sketchpad," 130–135.

42. Ibid., 132.

43. Manovich, *Software Takes Command*, 86.

44. Nanna Verhoeff, *Mobile Screens: The Visual Regime of Navigation* (Amsterdam: Amsterdam University Press, 2012), 66.

45. German Lopez, "Pokémon Go, Explained," Vox, August 5, 2016, http://www .vox.com/2016/7/11/12129162/pokemon-go-android-ios-game (accessed January 7, 2017).

46. The author observed this activity on a Montreal municipal bus in fall 2016.

47. Malcolm McCullough, *Ambient Commons: Attention in the Age of Embodied Information* (Cambridge, MA: MIT Press, 2013), 13–15.

48. Wendy Hui Kyong Chun, *Updating to Remain the Same: Habitual New Media* (Cambridge, MA: MIT Press, 2016), 171 and 53.

49. Apple, "Designing for iOS," https://developer.apple.com/library/iad/documenta tion/UserExperience/Conceptual/MobileHIG/index.html (accessed August 14, 2016). Contrast this with the similar effect of Microsoft's AERO operating system aesthetic, which Microsoft likens to glass, rather than paper.

50. Heidi Rae Cooley, "It's All about the Fit: The Hand, the Mobile Screenic Device and Tactile Vision," *Journal of Visual Culture* 3, no. 2 (August 2004): 137.

51. Jon Agar, *Constant Touch: A Global History of the Mobile Phone*, 2nd ed. (London: Icon, [2004] 2013), 180–181.

52. Apple, "iOS Human Interface Guidelines," https://developer.apple.com/library/ iad/documentation/UserExperience/Conceptual/MobileHIG/InteractivityInput .html (accessed August 14, 2016). This sense of intimacy generated sexist parodies upon the release of Apple's iPad in 2010. Apple's promotion of the iPad as small, slim, and easily toted in a bag provoked jokes comparing the device and its market-ing to maxi pads and other women's hygiene products. See Dawn Chmielewski and Alex Pham, "Women Mock the iPad Calling It the iTampon," *Los Angeles Times*, January 27, 2010, http://latimesblogs.latimes.com/technology/2010/01/women -mock-the-ipad-calling-it-itampon.html (accessed January 7, 2017); "Slogan Show-down: Tampons vs. iPad," *The Week*, January 28, 2010, http://theweek.com/ articles/497214/slogan-showdown-tampons-vs-ipad (accessed January 7, 2017).

53. Ingrid Richardson, "Pocket Technospaces: The Bodily Incorporation of Mobile Media," *Continuum* 21, no. 2 (June 2007): 214.

54. Ingrid Richardson, "Faces, Interfaces, Screens: Relational Ontologies of Framing, Attention and Distraction," *Transformations*, no. 18 (2010), http://www .transformationsjournal.org/wp-content/uploads/2017/01/Richardson_Trans18.pdf (accessed Janaury 7, 2017).

55. Regarding this understanding of interface, consider Branden Hookway's claim that "a human-machine interface, for example, would be fully bounded by the 'faces' of human and machine." Hookway, *Interface*, 9.

56. Lee Rainie and Kathryn Zickuhr, "Americans' Views on Mobile Etiquette," Pew Research Center, August 26, 2015, http://www.pewinternet.org/2015/08/26/americans-views-on-mobile-etiquette/ (accessed January 7, 2017).

57. Henrik Kaare Nielsen, "The Net Interface and the Public Sphere," in *Interface Criticism: Aesthetics Beyond Buttons*, ed. Christian Ulrik Andersen and Søren Bro Pold (Aarhus: Aarhus University Press, 2011), 218.

58. Jon Agar considers the many restrictions on phone use in public spaces, claiming they produce "oases of quiet." See Agar, *Constant Touch*, 241–252.

59. Jack Bratich and Heidi Brush, "Fabricating Activism: Craft-Work, Popular Culture, Gender," *Utopian Studies* 22, no. 2 (2011): 237. Emphasis in original.

60. Stella Minahan and Julie Wolfram Cox, "Stitch'nBitch: Cyberfeminism, a Third Place and the New Materiality," *Journal of Material Culture* 12, no. 1 (2007): 10–11. Emphasis in original.

61. Bratich and Brush, "Fabricating Activism," 236.

62. Kate Kershner, "Is It Rude to Knit in Public?" Howstuffworks, http://people.howstuffworks.com/is-it-rude-to-knit-in-public.htm (accessed January 7, 2017).

63. Judith Martin, "Miss Manners: Not Every Rite of Passage Is Supposed to Be Fun," *The Spokesman-Review*, August 1, 2011, http://www.spokesman.com/stories/2011/aug/01/not-every-rite-of-passage-is-supposed-to-be-fun/ (accessed January 7, 2017).

64. Jack Bratich, "The Digital Touch: Craft-Work as Immaterial Labour and Ontological Accumulation," *Ephemera* 10, no. 3–4 (2010): 308.

65. Bratich, "The Digital Touch," 308.

66. Rob Walker, "The D.I.Y. Revolutionaries of the Pussyhat Project," *The New Yorker*, January 25, 2017, http://www.newyorker.com/culture/culture-desk/the-d-i-y-revolutionaries-of-the-pussyhat-project (accessed April 15, 2017); see also https://www.pussyhatproject.com.

67. Constance Classen, "Feminine Tactics: Crafting an Alternative Aesthetics in the Eighteenth and Nineteenth Centuries," in *The Book of Touch*, ed. Constance Classen (Oxford: Berg, 2005), 229.

68. Matt Honan, "I, Glasshole: My Year with Google Glass," *Wired*, December 30, 2013, https://www.wired.com/2013/12/glasshole/ (accessed January 7, 2017).

69. Howard Rheingold, *Virtual Reality* (New York: Summit Books, 1991), 46.

70. Haraway, "Situated Knowledges," 189.

71. Anne Balsamo, *Technologies of the Gendered Body: Reading Cyborg Women* (Durham, NC: Duke University Press, 1995), 127.

72. Ken Hillis, *Digital Sensations: Space, Identity, and Embodiment in Virtual Reality* (Minneapolis: University of Minnesota Press, 1999), xv–xvi.

73. Ibid., xx.

74. J. David Bolter and Richard Grusin, *Remediation: Understanding New Media* (Cambridge, MA: MIT Press, 1999), 53 and 161.

75. Ibid., 54–55.

76. Bolter and Grusin further explain: "Virtual reality is a powerful expression of the desire for transparent immediacy and an enactment of the traditional male gaze that has been associated by Mulvey and others with the apparatus of the cinema." Ibid., 240 and 248.

77. Diane Gromala, "Pain and Subjectivity in Virtual Reality," in *Clicking In: Hot Links to Digital Culture*, ed. Lynn Hershman Leeson (Seattle: Bay Press, 1996), 222.

78. Ibid., 226.

79. Ibid., 224.

80. Cheris Kramarae, "Backstage Critique of Virtual Reality," in *CyberSociety: Computer-Mediated Communication and Community*, ed. Steve Jones (Thousand Oaks, CA: SAGE, 1995), 40.

81. Ibid., 37 and 45.

82. While Glass is no longer available to consumers, Google—now housed in the holding company Alphabet—has continued research into interactive eyewear and other wearables with remnants of the Glass development team working under the title "Project Aura." Alistair Barr, "Google Glass Gets a New Name and Hires from Amazon," *Wall Street Journal*, September 16, 2015, http://blogs.wsj.com/digits/2015/09/16/google-glass-gets-a-new-name-and-hires-from-amazon/ (accessed January 7, 2017).

83. Google, "Today We Announced +Project Glass," Google+, April 4, 2012, https://plus.google.com/+google/posts/YZYCe65gj8T (accessed January 7, 2017). My emphasis.

84. See especially Aruna D'Souza and Tom McDonough, eds., *The Invisible* Flâneuse*?: Gender, Public Space, and Visual Culture in Nineteenth-Century Paris* (Manchester: Manchester University Press, 2006).

85. Griselda Pollock, *Vision and Difference: Femininity, Feminism, and the Histories of Art* (New York: Routledge, 1988), 71. For more on the unraveling of male dominance through the flâneur and the crisis of gaze in modernity, see Elizabeth Wilson, "The Invisible Flâneur," *New Left Review* 191 (January–February 1992): 90–110; and David B. Clarke, *The Consumer Society and the Postmodern City* (New York: Routledge, 2003), 84–85.

86. Pollock, *Vision and Difference*, 67.

87. Google, "Glass: Explorers," https://sites.google.com/site/glasscomms/glass -explorers (accessed January 7, 2017).

88. Sergey Brin, "Sergey Brin Talks about Google Glass at TED 2013," TED, Long Beach, CA, 2013, https://www.youtube.com/watch?v=rie-hPVJ7Sw (accessed January 7, 2017).

89. Ibid.

90. Isabel Pedersen, *Ready to Wear: A Rhetoric of Wearable Computers and Reality-Shifting Media* (Anderson, SC: Parlor Press, 2013), 4–5. Emphasis in original.

91. Of course, eyeglasses have long been seen as masculine, as reflected in the saying "Boys don't make passes at girls who wear glasses."

92. Google, "Glass: Explorers."

93. "Google Cardboard," https://vr.google.com/cardboard/ (accessed August 14, 2016).

94. Sumit Passary, "Samsung, LG, Xiaomi and Others to Make Android Daydream VR-Ready Smartphones," *Tech Times*, May 20, 2016, http://www.techtimes.com/ articles/159512/20160520/samsung-lg-xiaomi-and-others-to-make-android -daydream-vr-ready-smartphones.htm (accessed January 7, 2017).

95. David Pierce, "The Future of Virtual Reality Is inside Your Smartphone," *Wired*, March 6, 2015, https://www.wired.com/2015/03/future-virtual-reality-inside -smartphone/ (accessed January 7, 2017).

96. Sherry Turkle, *The Second Self: Computers and the Human Spirit* (Cambridge, MA: MIT Press, [1984] 2005), 101–102.

97. Judy Wajcman, *Feminism Confronts Technology* (Cambridge: Polity, 1991), 156.

98. See Norman Bryson, "The Gaze in the Expanded Field," in *Vision and Visuality*, ed. Hal Foster (New York: The New Press, 1988), 88–94.

# Index

Adobe PhotoDeluxe, 76

Adobe Photoshop Elements, 76

Advertising, 73, 92, 104–105

Affect, 20, 56, 82, 86–93, 97, 101–106, 112–113. *See also* Labor, affective

Algorithms, 18–20, 38, 59, 75, 78, 87, 89, 91

Amazon, 167n6

Ambient commons, 10, 129

AMC Theaters, 122–123

American Civil War, 102

American Indian Movement, 52

America Online. *See* AOL

Anable, Aubrey, 92–93

Analytical Engine, 17, 24, 28, 30–34, 36–37, 61, 64

Andersen, Christian Ulrik, 8–9

Android (operating system), 93

AN/FSQ-7, 45, 47

AOL, 112

Apollo space program, 18, 49–51

Apple, 3, 12, 130–131

Applications. *See* Apps

Apps, 67, 78, 81, 86, 94, 104, 144
  craft, 6, 112–113
  game, 20, 86, 89, 91–94, 96, 99, 128
  and gender, 3
  locative, 75, 77, 128
  photography, 75
  photosharing, 82–84, 101
  settings, 67, 70

social media, 12, 20, 78, 82–84, 90
  writing, 96
  virtual reality, 138

AR. *See* Augmented reality

Aspray, William, 39

Assembly houses, 11, 18, 54–55

Attention, 10, 79, 92, 139
  and image, 69
  and interface, 129–132
  monetization of, 107
  and needlecraft, 133
  and virtual reality, 130, 136, 138

Augmented reality (AR), 21, 119, 135

"Aunt Bina's Quilt" (story), 102–103

Automation, 8, 155n61
  of electronics industry, 47, 49, 51, 53, 55
  of textile industry, 18, 24, 26, 28, 35–36, 152n24, 152n29

Avant-gardes, 13–14

Babbage, Charles, 17, 24, 28, 30–34, 36, 61–62, 64, 97

Barrett, Cyril, 170–171n45

Barrett, Edward, 13

Batchen, Geoffrey, 64, 72–73

Battin, Richard, 51

Baudelaire, Charles, 139

Bayard, Hippolyte, 61, 64

Bejeweled (game), 93

Benberry, Cuesta Ray, 102–103

Benjamin, Walter, 139
Bentley, Frank, 13
Betaworks, 94, 96
Binary code. *See* Code
Birmingham (England), 62
Blackshaw, Tony, 105
Body, 10, 121, 136, 144. *See also* Gender,
  and body; Hardware, and body;
  Interfaces, and body; Touch; Vision
  as computing metaphor, 30–31
  and industrial production, 56
  and mind, 20, 121, 124, 136–139
  and tactility, 131
Bolter, Jay David, 2, 137–138
Boris, Eileen, 108
*Boston Globe, The,* 133
Brackman, Barbara, 99
Bratich, Jack, 5, 16, 103, 112–113,
  133–135
Bridle, James, 6
Brin, Sergey, 119, 140–141
British Association for the Advancement
  of Science, 62
Brush, Heidi, 5, 16, 133
Bush, Vannevar, 61
Butler, Judith, 10–11
Byron, Anne Isabella, 33, 62
Byron, George Gordon, 35, 152n25

Calleja, Gordon, 93–94
Calvert, Melodie, 56–57
Cameras, 61–62, 64, 71–72, 118, 131,
  142
Campbell-Kelly, Martin, 39
Candy Crush Saga (game), 92–94
Cargoh, 5
Carpenter, Hilda G., 40–42, 47
Cartesianism, 72, 121
Cavell, Stanley, 68–69
CD-ROMs, 137
CGI, 6, 15, 77. *See also* Image quilting;
  Image stitching
Chat rooms, 15, 110, 124

China, 11, 53–54. *See also* Hong Kong
Chun, Wendy Hui Kyong, 8, 11–12, 30,
  57, 85, 129
Cinema, 19, 68–69, 80, 117–118, 132,
  138
Circuit boards, 18, 23, 51, 53, 55, 57,
  108
Circuits. *See* Integrated circuits
Classen, Constance, 120, 123, 135
Code, 15, 51, 57–58, 69–70
  binary, 18, 23, 31, 59, 71
  and Jacquard apparatus, 24
  and textile metaphors, 4
Coding, 4, 11, 13, 58, 72, 107, 112
Cold War, 38, 176n37
Colonialism, 49, 52, 99, 120, 123, 140
Computer-generated imagery. *See* CGI
Computers (devices), 11, 18, 23, 28,
  37–40, 51, 58, 74, 116, 143.
  description of, 30–31
  desktop, 12, 20, 51, 115
  laptop, 20, 23, 115, 124
  mainframe, 4, 31, 35, 49, 58, 124 (*see
  also* AN/FSQ-7; EDVAC; ENIAC; IBM
  704; SWAC)
  mini, 47
Computers (humans), 37–38, 153n34
Cooley, Heidi Rae, 131
Core memory, 39, 57–58. *See also* Rope
  memory
  assembly, 18, 40–45, 47–49, 52–53, 93
  configuration, 39–40, 43–44
  cost of, 39, 48–49
  decline of, 47, 51
  functioning of, 39
Cores. *See* Magnetic cores
Corporeality. *See* Body
Couchot, Edmond, 71–72
Courtauld silk mill, 36
Cowan, Ruth Schwartz, 106
Cox, Geoff, 107
Cox, Julie Wolfram, 5, 16, 133
Crabb, Carol Pinney, 102–103

Craftivism, 5, 133–135
Crafts, 4–7, 15–16, 43–45, 93, 111, 133–135
Craftsmanship, 15
*Critique of Judgment*, 67–68
Cyberfeminism, 16, 109–110, 123–124, 150n11
"Cyborg Manifesto, A," 109–110
Cyborgs, 90, 109–110

Daguerre, Louis-Jacques-Mandé, 61
Danet, Brenda, 15–16
Dangler, Jamie, 108
Daniels, Cynthia, 108
Darning. *See* Sewing
Data gloves, 137–138
de Beauvoir, Simone, 10
de Lauretis, Teresa, 120, 122
Derrida, Jacques, 67–69
Design, 2–3, 5, 11, 16, 31, 81, 116
    of apps, 79–84, 87, 99, 112
    of computers, 24, 30–33, 37, 125
    of games, 86, 93–94, 96
    of hardware, 58, 116–117, 119, 142
    of integrated circuits, 52–53, 66
    of memory, 37–43, 47–51
    of mobile media, 12–13, 21, 58, 85, 89, 130, 139
    of textiles, 24, 28, 32, 99, 101–102
Diamond Mine (game). *See* Bejeweled
Difference engine, 28, 32, 61–62
Digital Equipment Corporation, 47
Distraction, 10, 20–21, 104, 116, 119, 139
    and games, 20, 92
    mobile device as, 129–130, 141
Doane, Mary Ann, 122
Domestic sphere, 4, 10, 106, 113, 119–120, 134–135, 140
Dots (game), 94–96
Drawboy, 26
Drawgirl. *See* Drawboy
Drucker, Johanna, 7–8

Easiweave handloom, 43, 46–47
Eastman Kodak, 61
EDVAC, 30
Edwards, Paul, 38–39
Edwards, Steve, 161n10
Electronic Discrete Variable Automatic Computer. *See* EDVAC
Electronic Numerical Integrator and Computer. *See* ENIAC
Electronics industry, 11, 17, 47–57, 108–109
Elson, Diane, 55
Email, 110–111
Embodiment. *See* Body
Embroidery, 6, 15, 43, 51, 58, 82, 95–96, 103
Emoji, 82–83
ENIAC, 37–38
*Esquire* 3, 127
E-textiles. *See* Smart fabrics
Ethnicity, 53, 87, 108, 122. *See also* Race
Etsy, 5

Fabriculture, 5, 112
Facebook, 86, 93, 96, 102, 105, 120, 136
Fadeyev, Dimitry, 79
Fairchild Camera and Instrument. *See* Fairchild Semiconductor
Fairchild Semiconductor, 52–53, 71
Feilding, Elisabeth Theresa, 161n13
Feminism. *See* Cyberfeminism
Flânerie, 139–140
Flickr, 60, 81, 101
Forrester, Jay, 39
Frames, 1, 13, 18, 139, 142. *See also* Quilting, frames
    core memory, 39–40, 45, 47, 49, 51
    handloom, 43–47, 60, 68–71, 97
    image, 68–69, 139
    picture, 67–69
    screen, 67–71, 83–84, 94, 115, 117–118, 123, 131–132, 141
    textile assembly, 35, 152n24

Fraser, W. Hamish, 63
Friedberg, Anne, 67
Fuller, Matthew, 12, 97

Galloway, Alexander, 8, 9, 12, 150n11
Games, 13, 75, 77, 89, 97, 104, 124–
    125, 128
  casual, 20, 86–87, 91–96
  and gender, 92–93
  hardcore, 92
  match 3, 93
  and senses, 130–131
  and virtual reality, 137
Gelernter, David, 12
Gender, 2–7, 16, 85. See also Games,
    and gender; Ideology, and gender
  and ability, 52, 54–55
  and body, 2–3, 9–11, 20, 90, 115–116,
    123–124, 137
  depictions of, 125, 127, 134, 139–140
  divisions, 2–3, 35, 112, 116, 119, 122–
    123, 143–144
  and labor, 17–20, 28, 35–38, 42, 49–
    58, 87, 97, 103–113, 134–135
  and networks, 123
  and the senses, 17, 20, 119–123,
    139–140
  as social construction, 10
  and virtual reality, 21, 137–139
Geolocativity. See Locativity
Geotagging, 128. See also Tagging
Gesturality, 1, 4, 17, 56, 58, 78, 85, 116,
    131
  and games, 93–97
GIF, 6, 74
Girley 1 (game), 125–127
Girley 2 (game), 125–127
Gisler, Priska, 124
Gitelman, Lisa, 3, 12
Glasgow (Scotland), 63–64, 66
Glassholes, 142
Globalization, 2, 5, 17, 78, 99, 107–110
  and electronics industry, 48–49, 51–57

Godey's Lady's Book, 99–102
Goffey, Andrew, 12, 97
Google (search engine), 79
Google Cardboard, 21, 120, 142–143
Google Glass, 21, 119, 135, 139–142,
    144
Google Images, 81
Google Street View, 77
GPS, 13, 75, 128. See also Locativity
Graphical user interface. See GUI
Greenhill, Anita, 124
Gregg, Melissa, 106
Grids, 19, 31
  circuit, 4, 23, 39–44, 47, 51, 58, 77
  and games, 93–94
  informational, 38, 78, 80–82, 84, 90,
    93, 131
  pixel, 59, 72, 82
  and quilts, 93
  raster, 18, 72
  sewing pattern, 6
  woven, 18, 31, 51, 82
Gromala, Diane, 2, 138
Grosz, Elizabeth, 10–11, 121
Grusin, Richard, 137–138
GUI, 19–20, 39, 60, 78–82, 89–90, 93–
    96, 127–130
Guru (website), 110

Half-tone printing, 65–66
Halstead (England), 36
Haraway, Donna, 109–110, 123, 137
Hardware, 3, 11, 31, 69, 87–88, 113,
    116, 119–120
  and body, 11, 85, 115–116, 131–132,
    135–142
  design, 3, 21, 29–30, 38–41, 49, 85,
    117, 124–127, 142
  maintenance of, 38, 117–118
  production of, 17–18, 41–43, 47–57,
    66
Harris, Jan, 107
Harvey, Alison, 92

Hashtags, 89–93, 97
Hayles, N. Katherine, 11
Head-mounted displays, 120, 137, 138, 142
Heath, Stephen, 122
Herschel, John, 61, 64
Higgins, Hannah, 31
Hillis, Ken, 137
HMD. *See* Head-mounted displays
Home appliances, 34–35, 55, 57
Honan, Matt, 135, 141
Hong Kong, 53
Hookway, Branden, 7–10, 12, 127
Hopkins, Albert, 49–50
House of Lords. *See* Parliament (U.K.)
Howstuffworks, 133
Hu, Tung-Hui, 12
Huhtamo, Erkki, 118
Hula Girl (game). *See* Girley 2
Hyperlinks. *See* Hypertext
Hypermedia, 137–138
Hypertext, 90–91

IBM, 18, 45, 47–49, 53, 93
IBM 704, 47
Ickis, Marguerite, 103
Ideology, 1, 3, 4, 9, 87, 97, 113, 116, 119, 144
  and gender, 20, 54, 106, 120
  and interface, 7–8
  neoliberal, 8, 58, 107
  and vision, 122, 129, 139
Image quilting, 19, 60, 76–77
Images, 4, 6, 17–19, 96, 99, 105, 117–123, 128, 133. *See also* Photography
  arrangement of, 78–82, 89–90, 139–140
  production of, 15–16, 71–77, 119, 126–127
  flexibility of, 59–60, 66–71, 82–84
  woven, 23, 26, 28, 32
Image stitching, 19, 60, 75–77, 120
Indexicality, 60, 72–74, 128

Industrial Revolution, 28, 35, 106
Infinite scroll, 19, 60, 75, 78–83, 85
Instagram, 60, 78–82, 100–101, 105
Instrumentation Laboratory, 49
Integrated circuit boards. *See* Circuit boards
Integrated circuits, 51–53, 55, 57, 71
  assembly of, 52–53, 71
Interactivity, 2–3, 15, 31, 37, 77, 82–85, 89–97, 124–127
  haptic, 17, 118–119
Interfaces, 2, 11–13, 17–23, 85–91, 124–131, 139–141. *See also* Ideology, and interface
  aesthetics of, 66–69, 78–82, 94–97, 130
  and body, 56–57, 115–121, 131–132, 135–137
  textiles as, 60, 64
  theories of, 7–10
Internet, 4, 15, 74–75, 88–89, 110–111, 115–116, 124, 136–137. *See also* Networks
Internet Relay Chat. *See* IRC
Intimacy, 4, 20, 106, 115, 119, 123, 131–132, 142
iOS, 93, 130
iPad, 94, 177n52
iPhone, 2–3, 55–56, 78, 85, 93–94, 131
  design of, 58, 96
iPhone girl, 54–56
IRC, 15–16, 89
Italy, 61

Jackson, Shelley, 90
Jacquard, Joseph Marie, 17, 24, 29, 35
Jacquard apparatus, 17–18, 24–28, 31–35, 57, 62, 64
Jaiku, 89
Japan, 48–49, 51, 53
Jenning, Betty Jean, 38

Joysticks, 118, 124, 131, 137, 175n29
JPEG, 6, 74

Kant, Immanuel, 67–68
King (company), 92–93
King, Augusta Ada, Countess of
    Lovelace. *See* Lovelace, Ada
Kingston (New York), 49
Knitting, 6, 35, 42, 74, 119, 133–135,
    170n45
Knitting in public (KIP), 133–134
Kodak. *See* Eastman Kodak
KnitPro, 6
Kramarae, Cheris, 138

Labor, 2–3, 7, 17–20, 28, 86, 88, 91,
    96, 104–105. *See also* Mobility,
    and labor; Networks, and labor;
    Piecework; Race, and labor;
    Zero-work
  affective, 20, 86–87, 89, 93, 97, 104,
    112–113
  collective, 5, 19–20, 34, 87, 91, 97–99,
    103
  domestic, 4, 15, 20, 37, 54, 57, 92,
    104–108, 112–113, 133–135
  immaterial, 16–17, 20, 84, 86–87, 106–
    107, 109, 112, 137
  industrial, 11, 17, 20, 26, 28, 30, 35–
    37, 47–57, 105–106, 108–109
  protests, 37 (*see also* Luddism)
  unpaid, 20, 86, 105–106, 108, 111,
    133
  unskilled, 15, 36–37, 54–55, 144
  and wages, 36–37, 52–53, 105–07, 109
Landow, George, 90
Lazzarato, Maurizio, 107
Leisure, 8–9, 20, 35, 85–86, 92, 103–107,
    109–111
Levidow, Les, 54
Light, Jennifer, 38
Light gun, 124–127
Light pen, 124, 127

Lily handloom, 43, 45
Lincoln Laboratory, 11, 18, 39–43, 45
Linear perspective, 120, 123, 144
Lippard, Lucy, 103–104
Lister, Martin, 84, 164n38
Listservs, 124
Locativity, 12–13, 75, 77, 87, 128, 139
Looms, 2, 23, 31, 59–60
  hand-, 1, 13, 18, 32, 43–46, 48, 63,
    94
  industrial, 17, 24–28, 32–33, 35–36,
    48, 63
  screens as, 18, 68–71
Lovelace, Ada, 16, 17, 24, 31–33, 62, 64,
    150n11
Lowell (Massachusetts), 37, 51
Lowell, Francis Cabot, 37
Luddism, 35–36
Lyon (France), 35, 152n23

Mackenzie, Adrian, 12
Magic Loom, 43
Magnetic core memory. *See* Core
    memory
Magnetic cores, 18, 39, 41–44, 47–49,
    51, 57
Malaysia, 11, 53–54, 57
Male gaze, 123, 125, 133, 138–140
Manchester (England), 35
Manovich, Lev, 13–14, 59, 66, 127
Manufacturing, 17–18, 24–37, 47–58,
    63–64
Marchak, Patricia, 54
Masculinity, 10, 13–14, 16, 93, 112, 119,
    125
  and vision, 21, 119–120, 123, 132–
    133, 136, 139–141
Massachusetts Institute of Technology.
    *See* MIT
McAdams, Mindy, 115
McCullough, Malcolm, 10, 15, 129
McLean, Alex, 107
McLuhan, Eric, 15

Memory (computer), 38–39, 57–58, 61, 82. *See also* Core memory; Rope memory
  conceptualization of, 30–31, 37
  configuration of, 39–40, 43–45, 47–49
  cost of, 48
  and human computers, 38
  manufacture of, 2, 18, 40–43, 47–55
  mercury delay line, 31, 38
  punched card, 17, 31
  vacuum tube, 31, 38
Memory Test Computer, 40
Menabrea, Luigi Federico, 31
Merleau-Ponty, Maurice, 121, 141
Metadata, 78, 80, 89, 105
Metz, Christian, 122
MGI PhotoVista, 76
Microchips. *See* Integrated circuits
Mill girls, 37, 51, 56
Minahan, Stella, 5, 16, 133
Mindell, David, 51
Minecraft, 6
Mirzoeff, Nicholas, 120–122
Miss Manners, 133–134
MIT, 11, 18, 37, 39–43, 49, 51
Mitchell, William J., 72–73, 164n38
Moberg, Patrick, 94–96
Mobile media, 2–4, 12–13, 19–20, 86–89, 105–109, 121–123, 128, 130–142
Mobility, 9, 11–16, 19–20, 56, 86, 104, 115, 119. *See also* Vision, and mobility
  and immobility, 128
  and labor, 105–107
  in virtual reality, 130, 136
Morocco, 55
Mice (devices), 15, 78, 115, 124, 128
Mulvey, Laura, 122

Nakamura, Lisa, 11, 52
NASA, 18, 51

Navajo, 52
Navas, Eduardo, 13–14
Needlecraft. *See* Embroidery; Needlepoint; Quilting; Sewing
Needlepoint, 1, 6, 15, 43, 96, 103
Nelson, Theodor H., 90–91
Neoliberalism. *See* Ideology, neoliberal; Networks, and neoliberalism
Networks, 4, 12, 68, 99, 105, 109, 123, 136
  digital, 1–2, 5, 71
  functioning of, 77
  and images, 59
  and labor, 20, 97, 109–112
  social, 87, 98, 111
  and spaces, 10, 135
  and neoliberalism, 104–105, 111
  visualization of, 19
New Aesthetic, 6
Newton (device), 12
New York (city), 73
New York (state), 108–109
Niantic (company), 128
Nickel, Douglas, 64
Nielsen, Henrik Kaare, 132
Norman, Don, 116

Oculus Rift, 21, 120, 136
"One Day" (video), 139–140, 144
*On the Economy of Machinery and Manufactures*, 36
Operating systems, 8, 67, 70, 89, 122. *See also* Android; iOS
Operators (computer), 38

Packet switching, 77–78
PalmPilot, 12
Papian, William, 39–40, 154n46
Parergon, 67–69
Paris, 139–140
Parliament (U.K.), 35–36
Partner, Simon, 53, 159n105
*Patchwork Girl*, 90

Patterns, 1, 6, 18, 23, 60, 63–64, 107, 117, 131
  block, 33–34
  circuit, 51, 66
  commercial, 34–35
  and computing, 17, 32–33, 38, 64, 69–70
  and digital imagery, 19, 77, 79–81, 84, 86
  embroidery, 95–96
  and games, 93, 95–96
  fabric, 76–77
  handloom 43, 45, 52
  and hashtags, 90–91, 97
  Jacquard weaving, 26, 28, 32–33
  lace, 62, 64
  memory weaving, 48–49, 51
  Navajo weaving, 52
  quilt, 79–81, 99–101
  screen, 18, 65–66
  visualization, 52
PDP-8, 47
Pearson, Ruth, 55
Pedersen, Isabel, 141
Peer-to-peer textiling, 5, 103
Perlow, Seth, 56
Perspective. *See* Linear perspective
*Peterson's Magazine*, 99
Pew Research Center, 132
*Phenomenology of Perception, The*, 121
Philippines, 53
Phone covers, 2–3
Photobucket, 81
Photography, 6, 55, 56, 68, 118, 127, 139
  analog, 19, 71–74
  daguerreotype process
  digital, 19, 59–60, 61, 71–84
  evidentiary value, 71–74
  invention of, 61–62
  post-, 71–73
  and printing, 65–66
  and textiles, 60, 62–65
Photocells, 125
Photo editing, 74, 128. *See also* Image quilting; Image stitching
Photolithography, 51, 66, 71
Photosharing, 14, 78, 81–84, 89, 91, 99 101–102, 110
Picasa, 101
Piecework, 20, 35, 60, 87, 104, 107–108
Pimentel, Ken, 116
Pinterest, 78
Pixels, 6, 18, 59, 70, 72, 75–76, 78, 82
Plant, Sadie, 16, 59
Playdots (company), 94
PNG, 6, 74
Pokémon, 6. *See also* Pokémon Go
Pokémon Go, 128
Pold, Søren Bro, 8–9
Pollock, Griselda, 139–140
PopCap Games, 93
Processors, 51
Programmers, 17, 38. *See also* Operators
Project Whirlwind, 39–42, 125
Public sphere, 9–10, 21, 87, 113, 118– 121, 132–134, 141
Pun, Ngai, 54–55
Punched cards, 17–18, 26, 30–34, 57, 64. *See also* Memory
Pussyhat Project, 134–135
Putnam, Tim, 35

Quilting, 3, 11, 15, 19, 74–75, 87, 90, 93, 97–99, 102–104, 112
  bees, 1, 15–16, 20, 87, 97–99, 102
  chain piecing in, 79
  frames, 1, 13, 18, 98
  improv piecing in, 79
Quilts, 11, 16, 19, 79, 81, 97–99, 101–104
  album, 19

autograph, 101–102
patchwork, 19, 80–81

Race, 9, 11, 118, 122, 137
  and labor, 18, 42, 52–54
Radio assembly. *See* Transistor
  assembly
Ragan, Ralph, 50
Rancière, Jacques, 120, 127, 132
Ray-o-Lite Rifle Range, 124–125
Raytheon Corporation, 49–50
Remixing, 13–14
Rheingold, Howard, 136
Richardson, Ingrid, 132
Ritchin, Fred, 73–75
Robertson, Kirsty, 5
Rope memory, 18, 49–51
Rosler, Martha, 110
Rothstein, Natalie, 33, 152n23
Rowe, David, 105
Rubinstein, Daniel, 78, 80–81
Rutland, David, 38

SAGE, 39, 125, 127, 175n31
Sasson, Steve, 61
Saturn V rocket, 52
Scott, Mrs. O. W., 102
Scrapbooking, 5, 145n7
Screens, 1–4, 6, 23, 39, 88, 90. *See also*
  Looms, screens as
  cathode-ray tube (CRT), 125, 127
  cinema, 117–118
  computer, 59, 116, 124
  as frame, 67–69
  and gaze, 20–21, 118, 122–124, 128–
    129, 131
  and images, 59–60, 66–72, 74, 77, 81,
    94, 123, 130, 137
  light-emitting diode (LED), 59, 65–66,
    117
  orientation of, 67, 78–80, 82, 85,
    87
  printing, 65–66

size of, 66, 80, 115–116, 119, 121, 132,
  135
surface of, 1, 65–69, 85, 91, 94, 96,
  117–118, 123–124, 130–131, 144
television, 117, 119
touch-, 56–58, 79, 86–91, 107, 115–
  120, 123, 128–144
Seamlessness, 23, 73, 76, 121
Search engines. *See* Google; Google
  Images
Seeburg Corporation, 125
Semi-Automatic Ground Environment
  air defense system. *See* SAGE
Semiconductors. *See* Integrated circuits
Senses. *See* Touch; Vision
Sewing, 19–20, 34, 49, 75–77, 90,
  93, 108, 113. *See also* Embroidery;
  Needlepoint; Sewing machine
  in public, 119, 133–135
  training in, 37, 55
Sewing machine, 34–35, 55, 57, 79, 84,
  106, 108
Sharpston, Michael, 55
Shaw, Robert, 97–98
Shenzhen (China), 54
Shiprock (New Mexico), 52
Shiprock semiconductor assembly
  plant, 52–53
Shneiderman, Ben, 85
Silicon chips. *See* Integrated circuits
Sight. *See* Vision
Singapore, 53
Sketchpad (program), 127–128
Sluis, Katrina, 78, 80–81
Smart fabrics, 7
Smartphones, 11, 80, 118, 136, 141. *See
  also* iPhone
Snapchat, 60, 78–79, 82–84
Snapkidz, 83–84
Social media, 1–2, 4, 12–14, 75, 79–84,
  96, 123, 133–135, 139
  production, 86–90, 97–99, 102–106,
    110–112

Software, 4, 15, 17, 32, 35, 49, 69, 78, 81, 113, 129. *See also* Algorithms; Code
 design, 3, 13, 21, 58, 85, 87–88, 112, 130
 functioning of, 8, 23, 38, 57, 59, 71, 77, 84
 and metaphor, 4, 57
 photo editing, 75–76
 and weaving, 28, 58–59
Sondheim, Alan, 57–58
South Korea, 53
Space. *See* Domestic sphere; Public sphere
Sprite Stitch, 6
Standards Western Automatic Computer. *See* SWAC
Staples, David, 106–107
Sterne, Jonathan, 12
Stitch'nBitch, 5, 16, 133
Stone, Allucquère Rosanne, 123
Strasser, Susan, 106, 108, 152n29, 159n105
Strauven, Wanda, 118
Sundén, Jenny, 90
Superficiality, 1, 23, 57–58, 67, 93, 121. *See also* Screens, surface of
 and body, 10
Sutherland, Ivan, 127
SWAC, 38
Switch fabrics, 77

Tablets, 1, 23, 66, 84–85, 116, 118, 131. *See also* iPad
Tactility. *See* Touch
Tagging, 19, 81, 89–91, 97, 104, 113
Taiwan, 53
Talbot, William Henry Fox, 19, 61–66
Tapestry (app), 96
Taylor, Paul, 107
Taylorism, 56
TED, 140–141

Teixeira, Kevin, 116
Telecommuting. *See* Teleworking
Telephony, 77, 155n61
Television, 69, 117, 119
Teleworking, 111
Terranova, Tiziana, 107, 111–112
Terry, Jennifer, 56–57
Textiles, 1–7, 16–18, 23, 26–33, 59–62, 77, 98–102
 images as, 69–71, 76–77
 and photography, 19, 62–66
3-D, 75, 120, 131, 142
Touch, 20, 57, 67, 85, 103–104, 117–118, 130–131
 and body, 121–122
 elimination of, 120, 128, 135, 139, 141–142
 and gender, 123
Touchpads, 78, 128
Transistor assembly, 53, 159n105
Transistor girls, 53–55
Transistors, 47
Transparency, 68, 73, 137–138, 141
Transparent media, 137–138
Tumblr, 60, 78
Turkle, Sherry, 143–144
Twitter, 60, 79, 90, 96

Upwork, 110
U.S. Air Force, 39. *See also* SAGE
U.S. Army, 37–38

van Dijk, Jan, 111
Verhoeff, Nanna, 128
Victoria (queen of England), 62
Virtual reality, 21, 116, 119, 120–121, 130, 135–138, 142–44
Vision, 21, 76, 116, 118–119, 121–122, 131, 137–139. *See also* Ideology, and vision; Male gaze; Masculinity, and vision
 field of, 127–128
 and Google Glass, 139

and mobility, 123, 129
and politics, 127
and surveillance, 125
in virtual reality, 137–138
Visuality, 60, 72–73, 121–122
von Neumann, John, 30–31
Voyeurism, 141–142
VR. *See* Virtual reality
VR gloves. *See* Data gloves
VR headsets. *See* Head-mounted displays

Wajcman, Judy, 143
Walkie-talkie, 131–132
Waltham (Massachusetts), 50
Wearable media, 3, 7, 21, 119, 135, 137,
    139–142
Weave-It handloom, 43, 45
Weavers, 15, 17, 26, 32, 35–36,
    47150n3
Weaving, 2–3, 23, 37, 49, 69–71, 74,
    80, 109. *See also* Core memory,
    assembly
  home, 18, 35–37, 43–47, 68
  industrial, 17–18, 24–28, 30–36,
    63–64
  Navajo, 52
Websites, 78–79, 81, 92. *See also* Apps
Weinberger, David, 88
Whirlwind (computer), 38–44, 125
Whirlwind II. *See* AN/FSQ-7
Whirlwind Project. *See* Project
    Whirlwind
Widrow, Bernard, 41–42
Wilkes, Maurice, 41, 154n49
Williams, Michael, 47
Witmark (website), 110
Women's March on Washington,
    134–135
Work. *See* Labor
World War II, 30, 37–38, 125
World Wide Knit in Public Day, 134
World Wide Web. *See* Internet

Young, Iris Marion, 10
*Youth's Companion*, 102
YouTube, 104

Zero-work, 106–107
Zip files, 4
Zuckerberg, Mark, 97